Math Mammoth Grade 7 Skills Review Workbook Answer Key

By Maria Miller

Copyright 2020-2025 Taina Maria Miller
ISBN 978-1-942715-77-1

2020 Edition

All rights reserved. No part of this book may be reproduced or transmitted in any form or by any means, electronic or mechanical, or by any information storage and retrieval system, without permission in writing from the author.

Copying permission: For having purchased this book, the copyright owner grants to the teacher-purchaser a limited permission to reproduce this material for use with his or her students. In other words, the teacher-purchaser MAY make copies of the pages, or an electronic copy of the PDF file, and provide them at no cost to the students he or she is actually teaching, but not to students of other teachers. This permission also extends to the spouse of the purchaser, for the purpose of providing copies for the children in the same family. Sharing the file with anyone else, whether via the Internet or other media, is strictly prohibited.

No permission is granted for resale of the material.

The copyright holder also grants permission to the purchaser to make electronic copies of the material for back-up purposes.

If you have other needs, such as licensing for a school or tutoring center, please contact the author at
https://www.MathMammoth.com/contact

Contents

	Work-sheet page	Answer key page
Chapter 1: The Language of Algebra		
Skills Review 1	9	5
Skills Review 2	10	5
Skills Review 3	11	6
Skills Review 4	12	6
Chapter 2: Integers		
Skills Review 5	13	7
Skills Review 6	14	7
Skills Review 7	15	8
Skills Review 8	16	9
Skills Review 9	17	9
Skills Review 10	18	10
Skills Review 11	19	10
Skills Review 12	20	11
Skills Review 13	21	11
Skills Review 14	22	12
Chapter 3: Solving One-Step Equations		
Skills Review 15	23	13
Skills Review 16	24	14
Skills Review 17	25	14
Skills Review 18	26	15
Skills Review 19	27	16
Chapter 4: Rational Numbers		
Skills Review 20	28	17
Skills Review 21	29	18
Skills Review 22	30	18
Skills Review 23	31	19
Skills Review 24	32	19
Skills Review 25	33	20
Skills Review 26	34	20
Skills Review 27	35	21
Skills Review 28	36	21
Chapter 5: Equations and Equalities		
Skills Review 29	37	22
Skills Review 30	38	22
Skills Review 31	39	23
Skills Review 32	40	23
Skills Review 33	41	24
Skills Review 34	42	25

	Work-sheet page	Answer key page
Chapter 5: Cont.		
Skills Review 35	43	25
Skills Review 36	44	26
Skills Review 37	45	27
Skills Review 38	46	28
Skills Review 39	47	29
Skills Review 40	48	29
Chapter 6: Ratios and Proportions		
Skills Review 41	49	30
Skills Review 42	51	30
Skills Review 43	52	31
Skills Review 44	53	32
Skills Review 45	54	32
Skills Review 46	55	33
Skills Review 47	56	34
Skills Review 48	57	34
Chapter 7: Percent		
Skills Review 49	58	36
Skills Review 50	59	37
Skills Review 51	60	37
Skills Review 52	61	38
Skills Review 53	62	38
Skills Review 54	63	39
Skills Review 55	64	39
Skills Review 56	65	40
Chapter 8: Geometry		
Skills Review 57	66	41
Skills Review 58	67	42
Skills Review 59	68	43
Skills Review 60	69	44
Skills Review 61	70	44
Skills Review 62	71	45
Skills Review 63	72	45
Skills Review 64	73	46
Skills Review 65	74	47
Skills Review 66	75	48
Skills Review 67	76	48
Skills Review 68	78	49
Skills Review 69	79	50
Skills Review 70	80	50

	Worksheet page	Answer key page

Chapter 9: Pythagorean Theorem

Skills Review 71	81	51
Skills Review 72	82	51
Skills Review 73	83	52
Skills Review 74	85	54

Chapter 10: Probability

Skills Review 75	86	55
Skills Review 76	87	56
Skills Review 77	88	56
Skills Review 78	89	57
Skills Review 79	90	57
Skills Review 80	91	58

Chapter 11: Statistics

Skills Review 81	92	59
Skills Review 82	93	59
Skills Review 83	94	60
Skills Review 84	95	60
Skills Review 85	96	61
Skills Review 86	97	61
Skills Review 87	98	62
Skills Review 88	99	63
Skills Review 89	100	63
Skills Review 90	101	64

Chapter 1: The Language of Algebra

Skills Review 1, p. 9

1. a. 0.027   b. 32   c. 0.0001

2. a. 1/12   b. 25   c. 2/3

3.

a. $3 \cdot \dfrac{6}{2} = 9$	b. $2 + \dfrac{24}{3+1} = 8$	c. $\dfrac{42}{6} + \dfrac{1}{5} = \dfrac{7}{1} + \dfrac{1}{5} = 7\dfrac{1}{5}$
d. $\dfrac{3}{6} \cdot 2 = 1$	e. $2 + \dfrac{24}{3} + 1 = 2 + \dfrac{8}{1} + 1 = 11$	f. $12 - 1 + \dfrac{2}{3} = 11\dfrac{2}{3}$

4. a. $y - x^2$ b. $(y - x)^2$ c. $\dfrac{x-5}{x^3}$

5. a. $\dfrac{3}{4}m$

   b. $0.75m$

   c. $m + \dfrac{3}{4}m = \dfrac{7}{4}m$ or $m + 0.75m = 1.75m$

Skills Review 2, p. 10

1. $\dfrac{4p}{5} = 2{,}600$ and $p - (1/5)p = 2{,}600$

2. The root is 4.

3. a. Yes.
   b. No. For example, when n is 2, $(10 - 2) - 1$ is not equal to $10 - (2 - 1)$.
   c. No. For example, when n is 4, $(10 - 4) + 1$ is not equal to $10 - (4 + 1)$.

4. a. $x \cdot 5 = 75$ or, better written as $5x = 75$. The solution is $x = 15$.

   b. $1 - n = \dfrac{2}{7}$ Solution: $n = \dfrac{5}{7}$

5.

a. $(0.5 + 0.5) \cdot (9 - 7)^4$	b. $(0.7 - 0.3) \cdot 4^2$	c. $(2 \cdot 6^2 - 70)^2$
$= 1 \cdot 2^4$	$= 0.4 \cdot 16$	$= (2 \cdot 36 - 70)^2$
$= 16$	$= 6.4$	$= 2^2 = 4$

6. a. *perimeter* $= 4 \cdot 2x$, which can be simplified to $8x$
   b. *earnings* $= 15x + 20y$

Skills Review 3, p. 11

1. a. $a^2 + b^2$
   b. $(s - 2)^2 - 10$
   c. $(x + 3) - 10$

2. a. **(ii)** $(3 \text{ cm})^2$ b. **(i)** 27 cm^3

3. a. $2a + 3n + 7$
   b. $10y^2$
   c. $3x^4y^2$

4.

a. $5 \cdot 1 - 1^2 = 4$	b. $3 \cdot 10 \cdot 5 - 2 + 2 \cdot 10$ $= 150 - 2 + 20 = 168$
c. $\dfrac{3^2 - 1}{3 - 1} = \dfrac{8}{2} = 4$	d. $\dfrac{5}{5 - 1} + 1 = \dfrac{5}{4} + 1 = 2\dfrac{1}{4}$

5.

a. $2 \cdot (4 - 2)^2 \cdot (200 - 100)^2$ $= 2 \cdot 2^2 \cdot 100^2$ $= 2 \cdot 4 \cdot 10{,}000$ $= 80{,}000$	b. $9 + \dfrac{2 \cdot (5 - 3)^3}{2^3}$ $= 9 + \dfrac{2 \cdot 2^3}{2^3}$ $= 9 + 2 = 11$	c. $\dfrac{(1 + 1)^4}{4} + \dfrac{15 - 12}{2}$ $= \dfrac{2^4}{4} + \dfrac{3}{2}$ $= 4 + 1\tfrac{1}{2} = 5\tfrac{1}{2}$

6. a. is wrong. It should instead be r^3.
   b. is wrong. It should instead be $8x$.
   c. is correct.

Skills Review 4, p. 12

1. a. **(ii)** $(4x)^2$ b. **(iii)** $(4x)^3$

2. Area $= (4 \cdot 0.5 \text{ mi})^2 = (2 \text{ mi})^2 = 4 \text{ mi}^2$

3. $4(p - 5)$

4. a. Yes.
   b. No. For example, when n is 6, $(6 + 10) - 5$ is not equal to $5 - (10 + 6)$.

5. a.

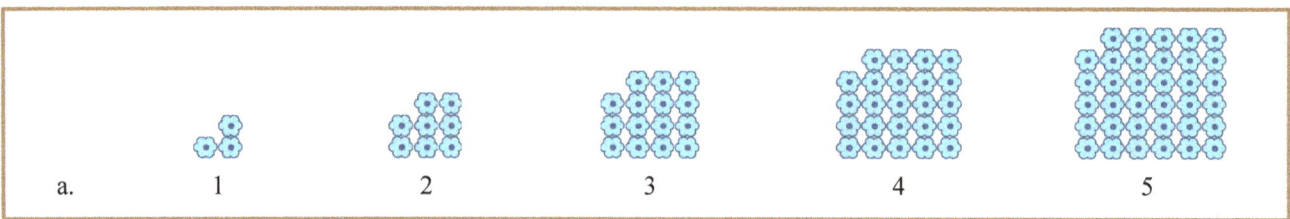

    a.   1   2   3   4   5

b. In each step, this pattern looks like a bigger and bigger square arrangement of flowers but there is always one flower missing from the upper left corner. Or, you could think of it growing by adding one flower to complete the square, and then also adding a row of flowers at the top and a column of flowers to the side.

c. 1,599 flowers

d. In step n, we have a square of the size $(n + 1)$ by $(n + 1)$ with one flower missing, so the formula is $(n + 1)^2 - 1$. Other possibility for the formula is: $n^2 + n + n$ which simplifies to $n^2 + 2n$.

Chapter 2: Integers

Skills Review 5, p. 13

1.
a. $60 \cdot (5-2)^2 - 20^2$
$= 60 \cdot 3^2 - 400$
$= 60 \cdot 9 - 400$
$= 140$

b. $11 + \dfrac{65-56}{2 \cdot (7-4)^3}$
$= 11 + \dfrac{9}{2 \cdot 3^3}$
$= 11 + \dfrac{1}{6}$
$= 11\dfrac{1}{6}$

c. $\dfrac{10^2}{(2+3)^2} \cdot \dfrac{23-8}{3}$
$= \dfrac{100}{25} \cdot \dfrac{15}{3}$
$= \dfrac{100}{5} \cdot \dfrac{3}{3}$
$= 20$

2. a. $7a - 77$ b. $200t - 200s - 100$ c. $0.4x + 0.8y + 2.8$

3. Each side is $\frac{1}{4} \cdot (52m + 8) = 13m + 2$.

4.

Expression	the terms in it	coefficient(s)	Constants
$2x^5$	$2x^5$	2	none
$7x + 14y + 8$	$7x$ and $14y$ and 8	7 and 14	8
$a + 0.97$	a and 0.97	1	0.97

5. a. $4w$
   b. $4w = 76$ m²; Solution: $w = 19$.

6.

a. $14x + 49 = \underline{7}(2x + \underline{7})$	b. $54y - 12 = 6(\underline{9y} - \underline{2})$
c. $36y + 45 = \underline{9}(4y + \underline{5})$	d. $108d + 60 = \underline{12}(9d + \underline{5})$

Skills Review 6, p. 14

1. a. $12m - 50$
   b. $m = (y + 50)/12$

2. a. The associative property of addition.
   b. The commutative property of multiplication.

3. a. 3   b. 11   c. −4   d. −9   e. −3

4. a. $100 - 6x = 46$
   b. Nine errors.

5.

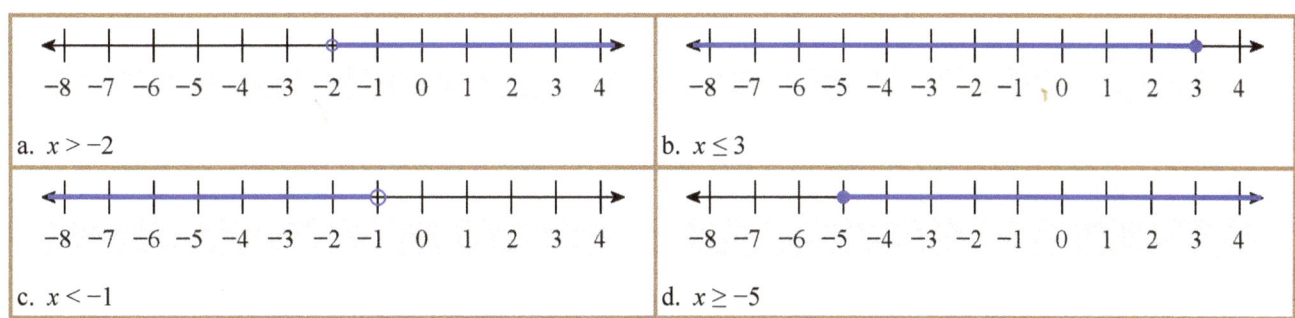

a. $x > -2$
b. $x \leq 3$
c. $x < -1$
d. $x \geq -5$

Skills Review 7, p. 15

1.

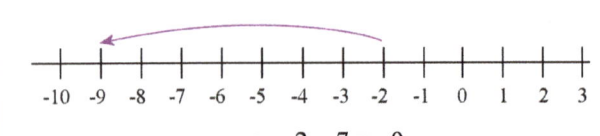
a. $-2 - 7 = -9$

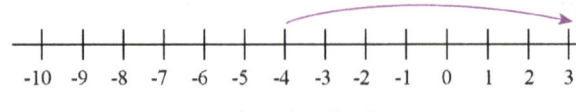
b. $-4 + 7 = 3$

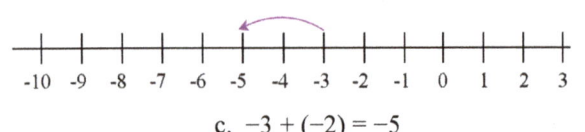
c. $-3 + (-2) = -5$

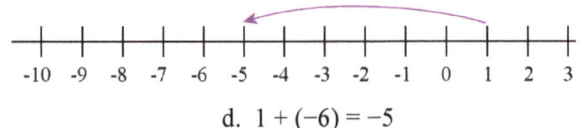
d. $1 + (-6) = -5$

2.

a. You are at ⁻5. You jump 9 to the right. You end up at **4**.

b. You are at ⁻1. You jump 10 to the left. You end up at **⁻11**.

Addition/subtraction:

⁻5 + 9 = 4

⁻1 − 10 = ⁻11

3. a.

b. 14m

c. 5m^2

4.

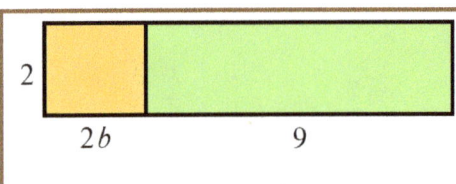
a. $2(\underline{2b} + \underline{9}) = 4b + 18$

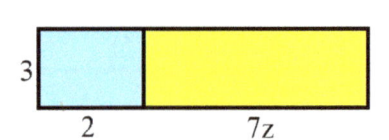
b. $3(\underline{2} + \underline{7z}) = 6 + 21z$

5.

a. $\dfrac{3^4}{3^2} \cdot (100 - 90)$

$= \dfrac{81}{9} \cdot 10$

$= 90$

b. $100 - \dfrac{(12 - 5)^2}{8 - 1}$

$= 100 - \dfrac{7^2}{7}$

$= 100 - 7 = 93$

Skills Review 8, p. 16

1.

a. $-7 + (-7) = -14$	b. $3 + (-6) = -3$	c. $50 + (-13) = 37$	d. $-13 + (-14) = -27$
$18 + (-7) = 11$	$-8 + 5 = -3$	$-40 + 14 = -26$	$10 + (-12) = -2$

2. a. On the number line, $-4 + 7$ is like starting at $\underline{-4}$, and moving $\underline{7}$ steps to the $\underline{\text{right}}$, ending at $\underline{3}$.
   b. With counters, $-4 + 7$ is like $\underline{4}$ negatives and $\underline{7}$ positives added together. We can form $\underline{4}$ negative-positive pairs that cancel each other out, and what is left is $\underline{3}$ positives.

3. Equation: $\$35 + 15p = \$80$. Solution: $p = \$3$.

4. a. $s \geq 120$ b. $t < -2$ c. $d \geq 50$ d. $m \leq 12$

5. a. $-|9| = -9$
   b. $-(6 - 2) = -4$
   c. $|-5| = 5$
   d. $-(4 + 8) = -12$

Puzzle Corner: $n = 1$ or $n = 0$.

Skills Review 9, p. 17

1.

a. $4 - (-11)$	b. $5 + (-2)$
$\downarrow$	$\downarrow$
$4 + 11 = 15$	$5 - 2 = 3$
c. $-6 - (-7)$	d. $4 + 9$
$\downarrow$	$\downarrow$
$-6 + 7 = 1$	$4 - (-9) = 13$

2. Yes, there is. Any of the negative numbers will work, because subtracting a negative number is the same as *adding* the opposite of the number, and thus the value of the expression becomes more than 5. Or, you can think of it as the double negative working out to be the same as a single plus sign. For example: $5 - (-1) = 6$, $5 - (-2) = 7$, and so on.

3. $5 - 4 - 3 - 4 = -6$

4. a.

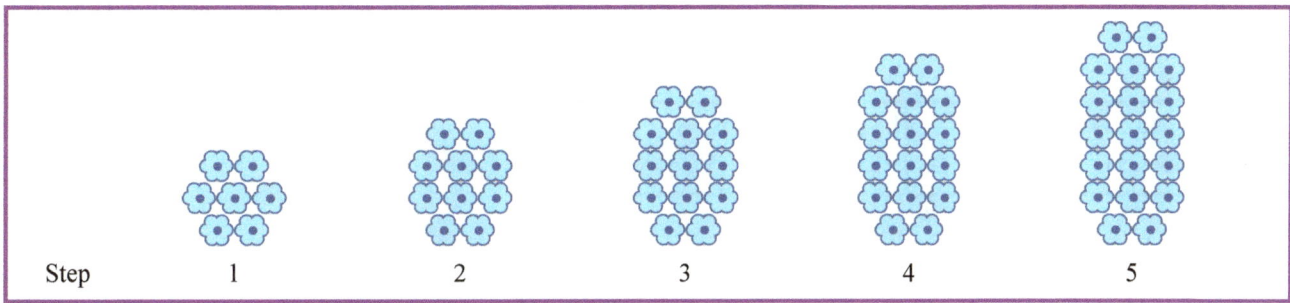

b. With each step, a row of three flowers is added to the pattern.
c. There will be 121 flowers in step 39.
d. There will be $3n + 4$ flowers in step n.

Skills Review 10, p. 18

1.

a. $\dfrac{800+1}{4} = \dfrac{800}{4} + \dfrac{1}{4} = 200\dfrac{1}{4}$	b. $\dfrac{30+1000}{5} = \dfrac{30}{5} + \dfrac{1000}{5} = 206$	c. $\dfrac{631}{6} = \dfrac{600}{6} + \dfrac{30}{6} + \dfrac{1}{6} = 105\dfrac{1}{6}$
d. $\dfrac{9x-3}{9} = \dfrac{9x}{9} - \dfrac{3}{9} = x - \dfrac{1}{3}$	e. $\dfrac{a+14}{7} = \dfrac{a}{7} + \dfrac{14}{7} = \dfrac{a}{7} + 2$	f. $\dfrac{4x+8}{8} = \dfrac{4x}{8} + \dfrac{8}{8} = \dfrac{x}{2} + 1$

2. a. *blue marbles* = $(1/3)m$ or $m/3$
   b. *not blue marbles* = $(2/3)m$ or $2m/3$
   c. *price* = $\$60 + (1/10) \cdot \$60$
   d. *price* = $(4/5)p$

3. *I'm tall when I'm young, and short when I'm old. What am I?*

−8		−15	−3	−15	−1	4	0
A		C	A	N	D	L	E

Puzzle Corner: $4 + n = 2 + 2 + n$ and $2n − 1 = n − 2 + n + 1$ are such equations.

Skills Review 11, p. 19

1. a.

a	6 + a	6 − a
−5	1	11
−4	2	10
−3	3	9
−2	4	8

a	6 + a	6 − a
−1	5	7
0	6	6
1	7	5
2	8	4

a	6 + a	6 − a
3	9	3
4	10	2
5	11	1
6	12	0

   b. When a is −5, −4, −3, −2, and −1, $6 − a$ is more than $6 + a$.
   c. When a is = −2.

2. a. Largest value when m is 11; smallest value when m is −6.
   b. Largest value when m is −6; smallest value when m is 11.

3. c. and e.

4.

a. a is −2 and b is 9	b. a is −1 and b is −8
$\lvert −2 − 9 \rvert = \lvert −11 \rvert = 11$	$\lvert −1 − (−8) \rvert = \lvert 7 \rvert = 7$

5. $0.75 \cdot 30m = 400$

Skills Review 12, p. 20

1. a. 7   b. −8   c. −8
   d. 7   e. 8   f. 8

2. a. −$50 − $12 = −$62
   b. −430 m + 80 m = −350 m
   c. 5°C − 7°C = −2°C

3. a. −14   b. 3   c. 57

4.

Operation	Commutative?	Associative?	An example
addition	yes	yes	3 + 6 = 6 + 3 and (6 + 2) + 7 = 6 + (2 + 7)
subtraction	no	no	6 − 2 ≠ 2 − 6 and (8 − 4) − 1 ≠ 8 − (4 − 1)
multiplication	yes	yes	5 · 8 = 8 · 5 and (3 · 9) · 4 = 3 · (9 · 4)
division	no	no	7 ÷ 5 ≠ 5 ÷ 7 and (6 ÷ 3) ÷ 2 ≠ 6 ÷ (3 ÷ 2)

5.

a. $8x + 2x = 10x$	b. $2y - 7$	c. $7m - 6n$
d. $2 - t$	e. $9s + s = 10s$	f. $u - u = 0$

Skills Review 13, p. 21

1. a. −20 − 10 + 15 = −15 dollars.
   b. −20 + 10 + 15 = 5 points.
   c. −20 + 10 − 15 = −25°C.

2.

a. 20 − (−3) − (−6) = 29	b. −9 − (−2) + (−3) = −10
c. −7 − 2 − (−3) + 4 = −2	d. 5 + (−3) − (−9) − 2 = 9

3.

a. $\dfrac{5^2}{5-1} = \dfrac{25}{4} = 6\dfrac{1}{4}$	b. $\dfrac{2 \cdot 3 - 1}{3^2} - 1 = \dfrac{5}{9} - 1 = -\dfrac{4}{9}$
c. $\dfrac{5 \cdot 2^2}{2+1} = \dfrac{5 \cdot 4}{3} = \dfrac{20}{3} = 6\dfrac{2}{3}$	d. $2 + \dfrac{9-0}{12-0} = 2 + \dfrac{9}{12} = 2\dfrac{3}{4}$

4. What belongs to you but other people use it more than you?

−30	−18	84	−12		−5	−8	24	2
Y	O	U	R		N	A	M	E

Skills Review 14, p. 22

1. b. and d.

2. a. $|-30-(-7)| = |-23| = 23$ b. $|30-(-7)| = |37| = 37$

3.

a. $6 \div (-4) = \dfrac{6}{-4} = -\dfrac{6}{4} = -1\dfrac{1}{2}$	b. $14 \div (-21) = \dfrac{14}{-21} = -\dfrac{14}{21} = -\dfrac{2}{3}$	c. $-32 \div 20 = \dfrac{-32}{20} = -\dfrac{8}{5} = -1\dfrac{3}{5}$
d. $-2 \div (-8) = \dfrac{-2}{-8} = \dfrac{1}{4}$	e. $-48 \div (-64) = \dfrac{-48}{-64} = \dfrac{12}{16} = \dfrac{3}{4}$	f. $12 \div (-96) = \dfrac{12}{-96} = -\dfrac{12}{96} = -\dfrac{1}{8}$

4. a. $-x > -6$ b. $|y| < 5$

5. a. $-(5+6) = -11$ b. $|3 \cdot (-4)| = 12$

6. $(\$4.50/500) \cdot s$ or $(\$4.50s)/500$ or $0.009s$

7. a. $-6x - 9$ b. $6n^3$ c. $-14s^2 - 3$

Puzzle corner:

a. $(9/10) \cdot 0.08p$ or $0.072p$ or $9 \cdot 0.08p/10$. There are other variations yet.

b. We can use an equation to solve this: $0.072p = 14$ from which $p = 14/0.072 = 194.444...$
She can print 194 pages.

Chapter 3: Solving One-Step Equations

Skills Review 15, p. 23

1.

a. $4 + 8 - 5 \cdot (-8)$ $= 4 + 8 - (-40)$ $= 4 + 8 + 40 = 52$	b. $(3 + (-4)) \cdot 9$ $= (-1) \cdot 9 = -9$	c. $4 - 5 \cdot 7 + 6$ $= 4 - 35 + 6$ $= -25$
d. $-1 + \dfrac{1}{2 - 5}$ $= -1 + \dfrac{1}{-3}$ $= -1 - \dfrac{1}{3}$ $= -1\dfrac{1}{3}$	e. $8 - \dfrac{1}{-4}$ $= 8 + \dfrac{1}{4}$ $= 8\dfrac{1}{4}$	f. $\dfrac{-12}{4} \cdot 2 + 9$ $= -3 \cdot 2 + 9$ $= -6 + 9 = 3$

2. a. $18(3x + 2)$
   b. $8(4x - 5y + 1)$
   c. $5(9a - 17b + 10)$
   d. $7(4 + 7a - 11b)$

3. a. $2(p + 300)$

   b. The final point count will be $2(-200 + 300) = 200$

   c. $2(p + 300) = -100$

   Solution: $p = -350$

4.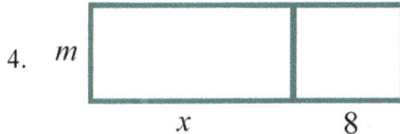

5.

a. $10 \cdot (-3) + 7 = -23$	b. $\dfrac{40}{-2} = 4 \cdot (-5)$	c. $14 - 3 \cdot 7 = -7$

Skills Review 16, p. 24

1. $|-15 - (-71)| = |56| = 56$.
   Roy has $56 more.

2. Yes, it is sometimes, but not always. When $n > m$, it will work. For example, the distance between 2 and 9 can be calculated as $9 - 2$, but the distance between 5 and 3 cannot be calculated as $3 - 5$.

3. a. See the values in the table.
   b. In August.

Months	High (C°)	Low (C°)	DIFFERENCE
January	-1	-11	10
February	2	-9	11
March	6	-4	10
April	12	2	10
May	18	7	11
June	23	12	11
July	26	15	11
August	26	14	12
September	21	10	11
October	15	4	11
November	9	-1	10
December	3	-7	10

4.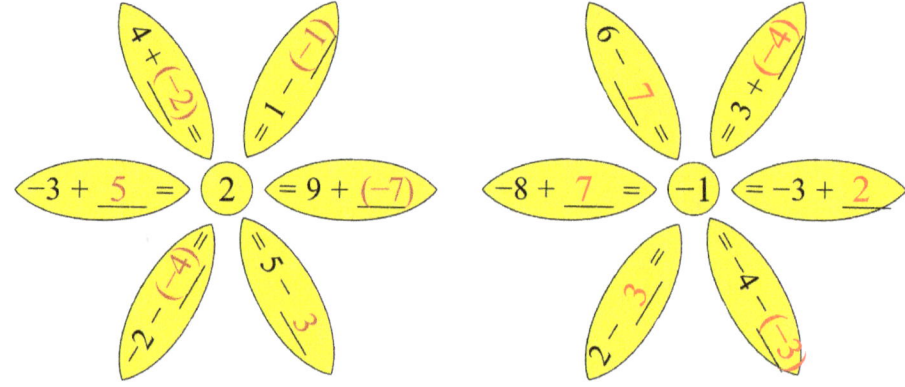

Skills Review 17, p. 25

1. a. $a - 8 = -1$. Solution: $a = 7$
   b. $6 = \dfrac{b+1}{3}$. Solution: $b = 17$
   c. $4(y - 1) = 12$. Solution: $y = 4$

2. a. 32   b. 60   c. −3,000   d. −56

3. $5x + 10y + 25s$

Skills Review 17, cont.

4.

Balance	Equation	Operation to do to both sides
$xx \oplus\oplus\oplus$ ⊖	$2x + 3 = -1$	-3
xx ⊖⊖⊖ ⊖	$2x = -4$	$\div 2$
x ⊖⊖	$x = -2$	

Skills Review 18, p. 26

1.

a.	-10	$=$	$x+6$	b.	-11	$= w - 20$
	-16	$=$	x		9	$= w$
	x	$=$	-16		w	$= 9$
c.	$-t$	$=$	$2 + (-5)$	d.	$8 - y$	$= -6$
	$-t$	$=$	-3		$-y$	$= -14$
	t	$=$	3		y	$= 14$

2. a. $3m - 7$ b. $7m^4$ c. $90y^4$

3. a. $18r^2$ b. $18r$

4.

5. Answers will vary. Check the student's answer. For example: $24 \div (-8) = -3$.

6. a. 13 b. -11 c. -5 d. 3

Skills Review 19, p. 27

1. a. $x = -120$ b. $x = 50$

2. a. $x = -81$ b. $s = 800$ c. $x = 33$

3. a. -40 b. 3   c. -11 d. 15

4. a.

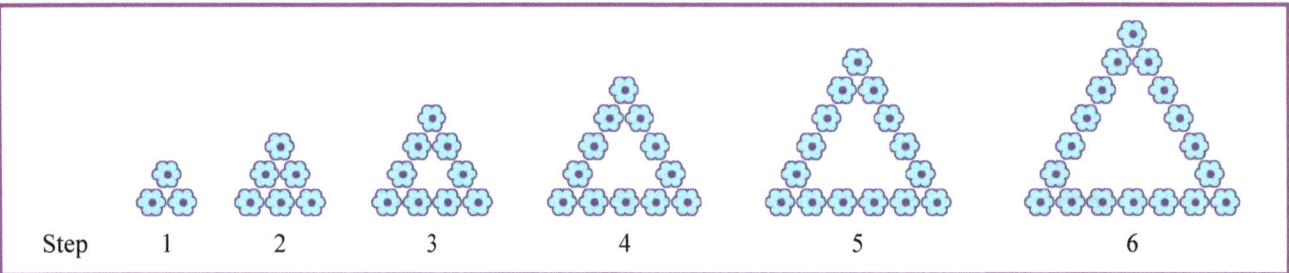

b. Answers will vary. For example: Each side of the triangle grows by one flower.
c. 117 flowers
d. $3n$ flowers

Chapter 4: Rational Numbers

Skills Review 20, p. 28

1. a. 11°C   b. 3 − (−8)

2. In part (a), the variable *p* denotes the number of pages, and in part (b), *w* denotes the length of the other piece, but naturally, other letters can be used for the variables.

   a. $20p = 1800$. Solution: $p = 90$

   b. $w + 3\ 1/4 = 8$. Solution: $w = 4\ 3/4$ (in feet).

3. a.

   b. Answers will vary. For example:

   $-11 \div 8 = \dfrac{-11}{8} = -1\dfrac{3}{8}$ or $44 \div (-32) = \dfrac{44}{-32} = \dfrac{11}{-8} = -1\dfrac{3}{8}$.

4.

| a. A Canadian goose flies a distance of 50 miles in 1 hour 15 minutes. What is his average speed?

$\begin{aligned} d &= v \cdot t \\ \downarrow \quad &\quad \downarrow \quad \downarrow \\ 50\text{ mi} &= v \cdot 75\text{ min} \\[4pt] 50\text{ mi} &= v \cdot 1.25\text{ hr} \\[4pt] \dfrac{50\text{ mi}}{1.25\text{ hr}} &= v \\[4pt] v &= \dfrac{50\text{ mi}}{1.25\text{ hr}} = 40\text{ mph} \end{aligned}$ | b. If a boy can bicycle at a constant speed of 9 km/h, how long will it take him to cover a distance of 15 km?

$\begin{aligned} d &= v \cdot t \\ \downarrow \quad &\quad \downarrow \quad \downarrow \\ 15\text{ km} &= 9\text{ km/h} \cdot t \\[4pt] \dfrac{15\text{ km}}{9\text{ km/h}} &= t \\[4pt] t &= \dfrac{15\text{ km}}{9\text{ km/h}} = 1\ 2/3\text{ hours} \\[4pt] &= 1\text{ hr }40\text{ min} \end{aligned}$ |

Skills Review 21, p. 29

1. a. (ii) and (iii)   b. s^3 c. 10 m

2. Answers will vary; check the student's answer. For example:

   "First, look at the plain numbers, or their absolute values, and subtract those. We get 714 − 293 = 421. Then check which of your original numbers had a greater absolute value - it is −714, and since that is negative, the answer also will be negative. So the final answer is −421."

3.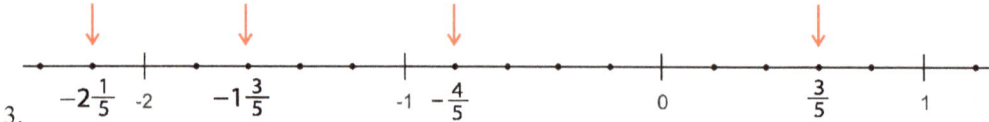

4. Answers will vary; check the student's answer. The two most commonly used commutative operations are addition and multiplication, and the ones that are not, subtraction and division. Numerical examples:

   Addition: 5 + 60 = 60 + 5 = 65

   Multiplication: 9 · 5 = 5 · 9 = 45

   Subtraction: 10 − 60 = −50 whereas 60 − 10 = 50.

   Division: 42 ÷ 6 = 7 whereas 6 ÷ 42 = 1/7.

5. Answers will vary; check the student's answer. Here are some examples:

   a. 3/1 or 6/2 or −3/(−1)   b. −87/1 or 87/(−1)   c. 74/100 or 37/50   d. −21/10 or 42/(−20)

6.

a. −8/6 = −4/3 = −1.$\overline{3}$	b. 10/(−25) = −2/5 = −0.4	c. −42/(−100) = 0.42

Skills Review 22, p. 30

1. 3(3x + 13) = 9x + 39

2. a. 3(10s − 1)   b. 6(7x + 6y + 8)

3. a. n · n − 2   b. For n = 11

4. a. Answers will vary; check the student's answer. For example: a cafeteria has a special where chicken enchiladas are discounted by $1.25, so you and your friends buy 10 servings.

   b. 10($x − $1.25) = 10x − $12.50

5.

a.			b.		
	8 =	−12 − x		11 − x =	7 − 13
	20 =	−x		11 − x =	−6
	−20 =	x		−x =	−17
	x =	−20		x =	17
c.	−5 + y =	−3 + 21	d.	2 + (−9) =	1 − z
	−5 + y =	18		−7 =	1 − z
	y =	23		−8 =	−z
				z =	8

Skills Review 23, p. 31

1. a. Yes, it is. Every repeating decimal is a rational number.

   b. $0.\overline{5}$ is more than 0.5. To find out how much more, we subtract:

   $$\begin{array}{r}0.555555...\\-0.5\\\hline 0.055555...\end{array}$$

   The number $0.\overline{5}$ is $0.0\overline{5}$ more than 0.5.

2. a. $3.58\overline{3}$ b. $0.3\overline{6}$ c. -0.05

3. a. $1/5 + (-7/8) = 8/40 + (-35/40) = \mathbf{-27/40}$ b. $-1/6 + 3/7 = -7/42 + 18/42 = \mathbf{11/42}$

4.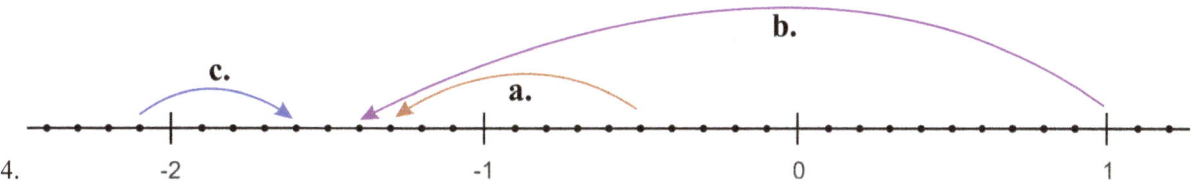

Skills Review 24, p. 32

1.

ratio	fraction	decimal	percent		ratio	fraction	decimal	percent
a. 11:25 =	$\frac{11}{25}$ =	0.44 =	44%		b. 3:50 =	$\frac{3}{50}$ =	0.06 =	6%

2. a. 0.01   b. -0.008 c. -0.0022
   d. -0.84 e. 0.024   f. 0.0025.

3. a. $-\dfrac{1}{26}$ b. $1\dfrac{5}{8}$ c. $-\dfrac{11}{54}$

4.

a. $-4 - (-11) - (-2)$	b. $9 - (-2) + (-7)$
$= -4 + 11 + 2$	$= 9 + 2 - 7$
$= 9$	$= 4$
c. $-10 - (-1) - (-3) + 3$	d. $4 + (-11) - (-7) - 1$
$= -10 + 1 + 3 + 3$	$= 4 - 11 + 7 - 1$
$= -3$	$= -1$

5. a. $7p$ b. $24p^3$ c. $48x^3y$

Skills Review 25, p. 33

1. a. −1.3348   b. 0.0014   c. 0.00064

2. a. 1 3/4   b. −9 3/5

3. a. In the third column, when x is zero, we cannot calculate the value of $3/x$ as it would lead to a division by zero.

x	$\frac{3}{x}$
6	$\frac{1}{2}$
5	$\frac{3}{5}$
4	$\frac{3}{4}$

x	$\frac{3}{x}$
3	1
2	$1\frac{1}{2}$
1	3

x	$\frac{3}{x}$
0	—
−1	−3
−2	$-1\frac{1}{2}$

x	$\frac{3}{x}$
−3	−1
−4	$-\frac{3}{4}$
−5	$-\frac{3}{5}$

b. $x = -6$

Skills Review 26, p. 34

1. a. $5.06 \cdot 10^5$ b. $3.4099 \cdot 10^7$
   c. $1.72 \cdot 10^4$ d. $6.445 \cdot 10^8$

2. a. $|x| < 6$
   Example values: $x = 2$ and $x = -1$

   b. $-x < 2$
   Example values: $x = 3$ and $x = -1$

   c. $-x > -5$
   Example values: $x = 2$ and $x = -7$

3. a. $x = -160$ b. $x = -900$ c. $x = -24$

4. Let t be the time Joan has bicycled up to now (the unknown). Equations will vary. Check the student's equation. For example:

   10 km/h · t + 30 km = 65 km   Or: 10 km/h · t = 35 km   Or: 35 km = 10 km/h · t

   Solution: $t = 3.5$ hours.

Skills Review 27, p. 35

1. a. $x = 1\,7/8$ b. $v = -9/10$

2. a. (3)   b. (2)   c. (3)

3. Answers will vary. Check the student's answer. For example:

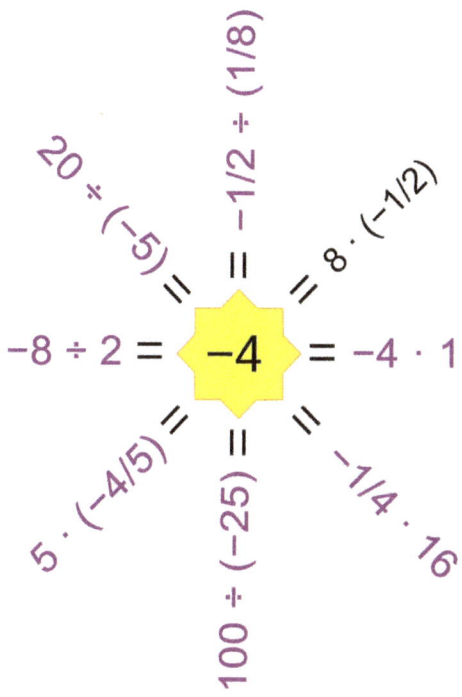

4. We can use the formula $d = vt$ but this can also be solved without an equation.

   Equation: $5.6 \text{ km} = 12 \text{ km/h} \cdot t$

   Solution: $t = 5.6 \text{ km}/(12 \text{ km/h}) = 0.4666...$ hours $= 28$ minutes.

Skills Review 28, p. 36

1. a. $2x + t$ b. $-9 + 2x$ c. $-5w$
   d. $-s$ e. $8 - 7x$ f. $7h - 10g$

2. a. $x = -3/32$ b. $z = -7\,1/3$

3.

a. $\dfrac{1}{3} - 1.2 - \dfrac{3}{4}$	b. $0.4 \cdot \dfrac{5}{6} - 2\dfrac{1}{2}$
$= \dfrac{1}{3} - \dfrac{6}{5} - \dfrac{3}{4}$	$= \dfrac{2}{5} \cdot \dfrac{5}{6} - 2\dfrac{1}{2}$
$= \dfrac{20 - 72 - 45}{60} = \dfrac{-97}{60} = -1\dfrac{37}{60}$	$= \dfrac{2}{6} - 2\dfrac{1}{2}$
	$= \dfrac{2}{6} - \dfrac{15}{6}$
	$= -\dfrac{13}{6} = -2\dfrac{1}{6}$

Chapter 5: Equations and Inequalities

Skills Review 29, p. 37

1.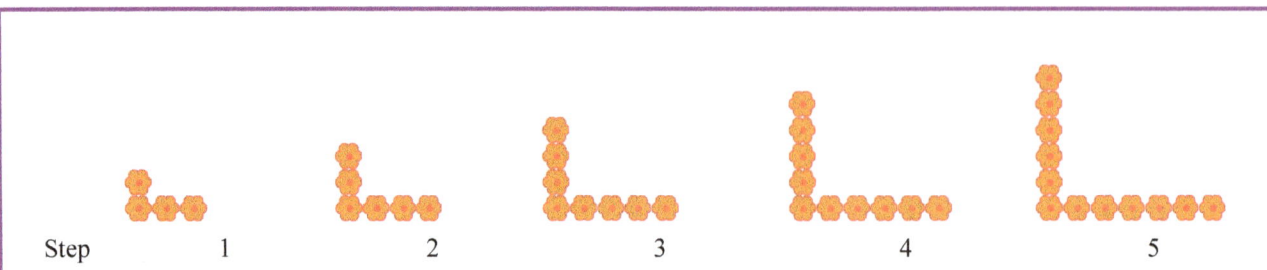

   a. See the pattern above.

   b. Answers will vary. Check the student's answer. For example:
   At each step, it adds one flower to the top and one to the right.

   c. 80 flowers

   d. Answers will vary. Check the student's answer. For example, you can see the pattern as having $n+1$ flowers standing up, and another $n+1$ flowers laying horizontally, thus a total of $2(n+1)$ flowers.

   Or, you can think of the step one having 4 flowers, and then after that, it adds 2 flowers in each step, so the generic formula is or $4 + 2(n-1)$.

2. a. $x = 0.9$ b. $y = 1.74$ c. $x = -17.2$ d. $x = 1.\overline{06}$

3.

| a. $5\frac{9}{10,000}$ | b. $-37\frac{3,920,483}{10,000,000}$ | c. $60\frac{605}{100,000}$ |

4. a. -0.0589 b. 205.5 c. 0.040954

Skills Review 30, p. 38

1. a. The side is unknown. Let s denote the unknown side. The equation is $4s = 5.6$ m or $s = 5.6$ m $/ 4$.

   b. Kyle's age is the unknown. Let K denote Kyle's age. Equation: $K - 15 = 56$ or $56 + 15 = K$.

2.

a.	$12 = -7 - x$	b.	$-45 + r = -15 + (-9)$
	$19 = -x$		$-45 + r = -24$
	$-19 = x$		$r = 21$
	$x = -19$		

3. c.

4. a. and c.

5.

| a. $6 + 3 \cdot (-5) = -9$ | b. $-24 \div 6 + 4 \cdot 1 = 0$ | c. $(-5 \cdot 4) + 8 + 2 = -10$ |

Skills Review 30, cont.

6.

a. $-\dfrac{1}{4} - \left(-\dfrac{1}{2}\right) + \dfrac{5}{8}$	b. $\dfrac{2}{3} + \left(-\dfrac{5}{6}\right) + \dfrac{1}{4} + \left(-\dfrac{1}{2}\right)$
$= -\dfrac{1}{4} + \dfrac{1}{2} + \dfrac{5}{8}$	$= \dfrac{2}{3} - \dfrac{5}{6} + \dfrac{1}{4} - \dfrac{1}{2}$
$= -\dfrac{2}{8} + \dfrac{4}{8} + \dfrac{5}{8}$	$= \dfrac{8}{12} - \dfrac{10}{12} + \dfrac{3}{12} - \dfrac{6}{12}$
$= \dfrac{7}{8}$	$= \dfrac{8 - 10 + 3 - 6}{12} = -\dfrac{5}{12}$

Skills Review 31, p. 39

1.

a.	$\dfrac{x-6}{7} = 5$	b.	$\dfrac{y+20}{-7} = 1$
	$x - 6 = 35$		$y + 20 = -7$
	$x = 41$		$y = -27$
c.	$9 - 3h = 3$	d.	$22 = 10 - 4t$
	$-3h = -6$		$12 = -4t$
	$h = 2$		$-3 = t$
			$t = -3$

2. a. 1,050 $^7/_8$ b. 1,200 $^3/_4$ c. 230 $^1/_3$
   d. $x - ½$ e. $7a + 1$ f. $-b + ½$

3. $2e$

4. a. 7.6   b. -1.52 c. -1.918

Skills Review 32, p. 40

1. a. The table below has eight values but the student is only required to find six.

x	$-6 - 2x$
-3	0
-2	-2
-1	-4
0	-6

x	$-6 - 2x$
1	-8
2	-10
3	-12
4	-14

b. For the value $x = -3$.

Skills Review 32, cont.

2. a. ⬭$p - \frac{p}{5} = \$12$⬭ $\frac{p}{5} = \$12$ $\frac{5p}{4} = \$12$ ⬭$\frac{4p}{5} = \$12$⬭ $p - 1/5 = \$12$

b. ⬭$\frac{3000 - 600}{75} = x$⬭ $\frac{3000 - x}{75} = 600$

⬭$75x + 600 = 3000$⬭ $75(x + 600) = 3000$

3. Answers will vary. Check the student's answer. For example:

a. $1.08 \cdot \$315 = \$340.20$

A sales tax of 8% was added to the price of a $315 lawn mower and the final price was $340.20.

b. $(9/10) \cdot 160 \text{ lb} = 144 \text{ lb}$

Sheila weighed 160 lb and lost 10% of her weight. Now she weighs 144 lb.

c. $(1/2) \cdot (3/4) = 3/8$

There was 3/4 of a chocolate bar left. Robert ate half of that. This means he ate 3/8 of the entire chocolate bar and now there is 3/8 of the bar left.

Skills Review 33, p. 41

1.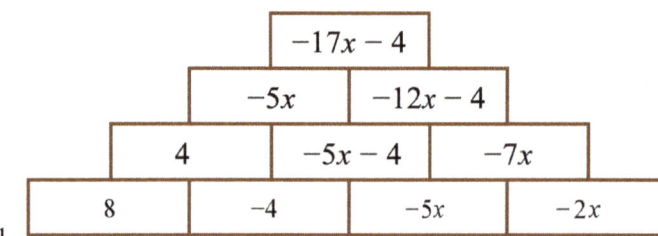

2.

a.
$-6x + 2x = 4x - x - 6$
$-4x = 3x - 6$
$-7x = -6$
$x = 6/7$

b.
$4 + 50y - 10y = 10 - 20y$
$4 + 40y = 10 - 20y$
$40y = 6 - 20y$
$60y = 6$
$y = 1/10$

3. Draw a bar model for the equation $12.6 + 0.9 + w + 91.9 = 290$, and solve it.

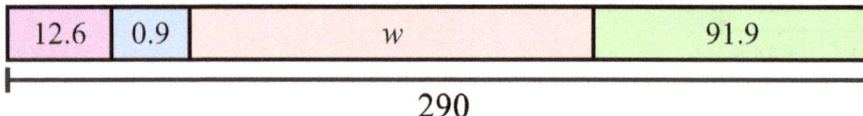

Solution: $w = 290 - 91.9 - 0.9 - 12.6 = \mathbf{184.6}$

Skills Review 33, cont.

4.

a. Yes, the expressions are equal.	b. No, they are not equal. For example, when n = 1, $\dfrac{5}{n+1} = \dfrac{5}{2}$ but $\dfrac{n+1}{5} = \dfrac{2}{5}$

Skills Review 34, p. 42

1. a. $3 + 2(n - 1)$ or $2n + 1$

   b. 57 flowers

   c. In row 48

2. $2x + 12 + 5y$
   (or $2x + 5y + 12$)

3. a. -8 b. -6 c. -40

4. a. $m \leq -2000$

   b. $g \geq 21$

   c. $t \leq 5°$

5. a. $x = -9/80$ b. $x = 2/15$

Row	Flowers
1	3
2	3 + 2
3	3 + 4
4	3 + 6
5	3 + 8
6	3 + 10
7	3 + 12
8	3 + 14
n	3 + 2(n − 1)

Skills Review 35, p. 43

1. a. The average is $\dfrac{7 + 2 - 4 - 7 - 6 - 2}{6} = \dfrac{-10}{6} = -1\dfrac{2}{3}°$.

   b. The average is $\dfrac{7 + 2 - 4 - 7 - 6 - 2}{6} = \dfrac{-10}{6} = -1.67°$.

2. a. $20t - 120$ b. $0.5a + b$ c. $21x + 14y + 2.8$

3.

a. $6 \cdot 99 = 6(100 - 1)$ $= 600 - 6 = 594$	b. $4 \cdot 98 = 4(100 - 2)$ $= 400 - 8 = 392$	c. $7 \cdot 299 = 7(300 - 1)$ $= 2{,}100 - 7 = 2{,}093$

4. a. $x = 48$ b. $x = -140$ c. $x = -16$ d. $z = 200$ e. $y = 135$

Skills Review 36, p. 44

1.

a.
$$3x - 22 < 32$$
$$3x < 54$$
$$x < 18$$

(number line showing x < 18, open circle at 18)

b.
$$-50 \leq 10x + 30$$
$$-80 \leq 10x$$
$$-8 \leq x$$
$$x \geq -8$$

(number line showing x ≥ −8, closed circle at −8)

2. Equation: $-11°C - 3° + 7° = -7°$
   Now the temperature is $-7°$.

3.

a. $-40 + (-13) - (-12)$ $= -40 - 13 + 12 = \mathbf{-41}$	b. $0 - (-2) - (-9)$ $= 0 + 2 + 9 = \mathbf{11}$
c. $-9 + (-8) + 10 + (-8)$ $= -9 - 8 + 10 - 8$ $= \mathbf{-15}$	d. $14 - (-21) - 7 - 3$ $= 14 + 21 - 7 - 3$ $= 35 - 7 - 3 = \mathbf{25}$

4. The first fifth of the journey is 14 miles. From the formula $d = vt$, we get

   $14 \text{ mi} = 35 \text{ mph} \cdot t_1$

   $t_1 = 14/35 \text{ hr} = 2/5 \text{ hr} = 24$ minutes

   The other part is 56 miles. From the formula $d = vt$, we get

   $56 \text{ mi} = 50 \text{ mph} \cdot t_2$

   $t_2 = 56/50 \text{ hr} = 1 \; 3/25 \text{ hr} = 1.12$ hours $= 1$ hr 7.2 minutes

   In total, it takes 24 minutes + 1 hr 7.2 minutes = 1 hr 31.2 minutes, or about 1 1/2 hours.

5.

a. $\dfrac{2}{\frac{5}{9}}$

$= 2 \div \dfrac{5}{9} = 2 \cdot \dfrac{9}{5} = \dfrac{18}{5}$

$= 3\dfrac{3}{5}$

b. $\dfrac{\frac{3}{8}}{\frac{1}{12}}$

$= \dfrac{3}{8} \div \dfrac{1}{12} = \dfrac{3}{8} \cdot 12 = \dfrac{36}{8}$

$= 4\dfrac{1}{2}$

c. $\dfrac{\frac{4}{5}}{-10}$

$= \dfrac{4}{5} \div (-10) = \dfrac{4}{5} \cdot \left(-\dfrac{1}{10}\right)$

$= -\dfrac{4}{50} = -\dfrac{2}{25}$

Skills Review 37, p. 45

1. Let *n* be the number of times Sheila can ride SuperFall. We can write the inequality:

    $2 \cdot \$5.50 + n \cdot \$3.50 \le \$28$

    You can solve this with logical reasoning and get the answer that she can ride it four times. Here is how to solve it with the inequality:

    $\$11 + \$3.50n \le \$28$

    $\$3.50n \le \$17$

    $n \le 4.857...$

    She can ride it four times.

2.

Three-eighths of a number is 12.45. What is the number?	
Equation: $\frac{3}{8}x = 12.45$ $\quad \cdot 8$ $3x = 99.6$ $\quad \div 3$ $x = 33.2$	Another way: 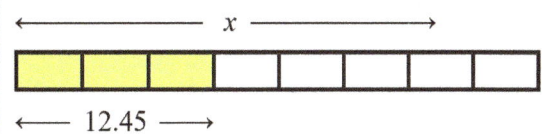 $\leftarrow$ 12.45 $\rightarrow$ One block is $12.45 \div 3 = 4.15$. Then, all eight blocks are $8 \cdot 4.15 = 33.2$

3. a. $560{,}000 = 5.6 \cdot 10^5$ b. $3{,}290{,}000 = 3.29 \cdot 10^6$
   c. $40{,}100 = 4.01 \cdot 10^4$ d. $24{,}500{,}000 = 2.45 \cdot 10^7$

4.

a. -2	6		b. 3	c. 2
1				4
	d. -5	6	0	
e. 4	0			e. -7
		f. -5	0	0

Skills Review 38, p. 46

1.

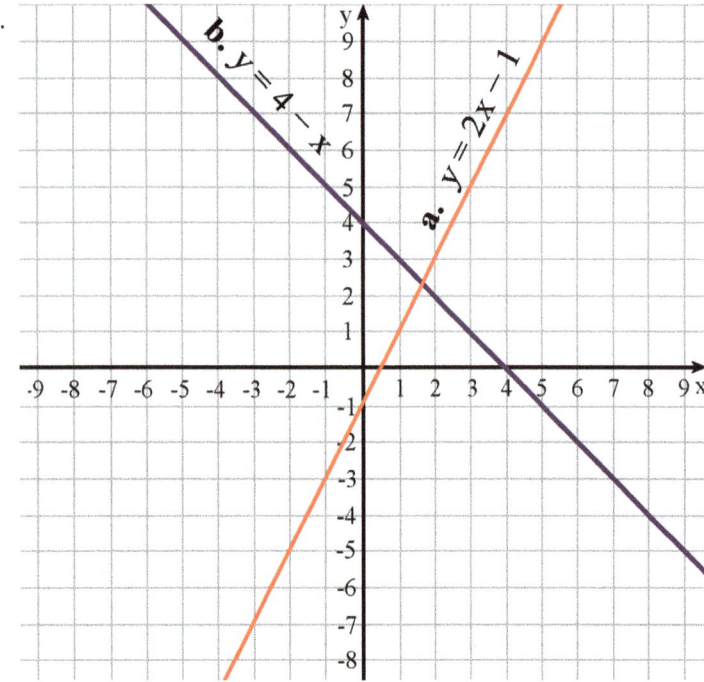

2. a. 2   b. −1

3. (1) We can substitute the coordinate values x = 2 and y = −3 to the equation of the line, and check if we get a true or false equation. The line is $y = 4 - x$ so we get: $-3 = 4 - 2$ or $-3 = 2$, which is a false equation. So the point is not on the line.

   (2) You can plot the line and the point, and check if the point is on the line visually.

4. Answers will vary. Check the student's answer. Here are three possibilities.

(1) $(-8 + 5) \cdot 3 - 12$	(2) $-8 + 5 \cdot (3 - 12)$	(3) $-8 + (5 \cdot 3 - 12)$
$= -3 \cdot 3 - 12$	$= -8 + 5 \cdot (-9)$	$= -8 + 3$
$= -21$	$= -8 + (-45)$	$= -5$
	$= -53$	

5.

a.	$31 = 5 - 2x$ $26 = -2x$ $-13 = x$ $x = -13$	b.	$3 - 4t = -13$ $-4t = -16$ $t = 4$
c.	$\frac{x}{5} + 3 = -5 \cdot (-3)$ $\frac{x}{5} + 3 = 15$ $\frac{x}{5} = 12$ $x = 60$	d.	$\frac{-3x}{4} = -21$ $-3x = -84$ $x = 28$

Skills Review 39, p. 47

1. a. $|x-y|$

   b. Sheila is $240 better off.

   c. $|\$40 - (-\$200)|$ or $|-\$200 - \$40|$
      (Both will work and give the answer of $240.)

2. a. $-7 + (-3) + 2 = -8$

   or $-11 + (-6) + 3 = -8$

   b. $5 + 6 + (-11) = 0$

   or $-7 + 5 + 2 = 0$

3. a. 1.53846
   b. $0.16\overline{3}$

4. See the image on the right.
   a. The slope is 1/2.
   b. The slope is -2.

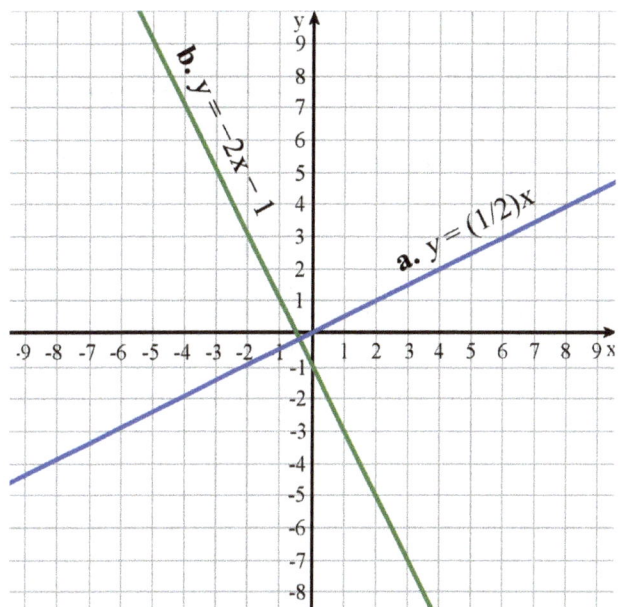

Skills Review 40, p. 48

1. The table on the right has eight values but the student is only required to find six.

   b. The expression $3x - 2$ has the value of 0 when $x = 2/3$.

x	3x − 2
−3	−11
−2	−8
−1	−5
0	−2

x	3x − 2
1	1
2	4
3	7
4	10

2. a.

t	0	1	2	3	4	5	6	7
d	0	8	16	24	32	40	48	56

   equation: $d = 8t$

   b.

t	0	1	2	3	4	5	6	7
d	0	9	18	27	36	45	54	63

   equation: $d = 9t$

   c. The second runner will be 10 meters ahead of the first when $t = 10$ seconds.

3. $420(2.25 + x) = 1050$
    $945 + 420x = 1050$
    $420x = 105$
    $x = 0.25$

They should increase the price by $0.25, to be $2.50.

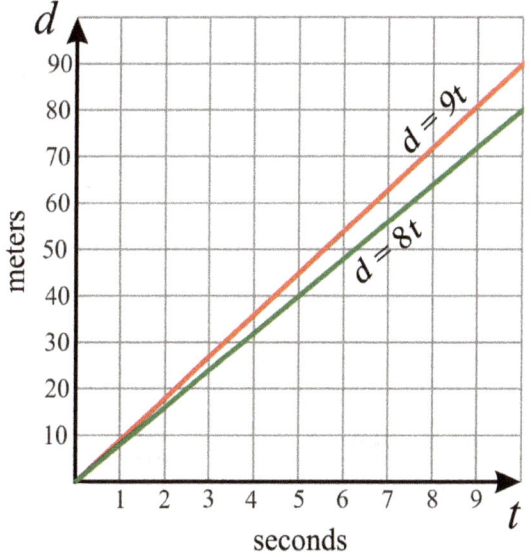

Chapter 6: Ratios and Proportions

Skills Review 41, p. 49

1. a. See the graph on the right.
   b. Henry's average speed during the first hour is 20 km/h.
      During the second hour it is also 20 km/h.
   c. $d = 20t$
   d. Andy's average speed during the first hour is 20 km/h.
      During the second hour it is 16 km/h.
   e. At 4 hours, Henry will be at 80 km, and Andy will be at 68 km.
      So, Henry will be 12 km further than Andy.

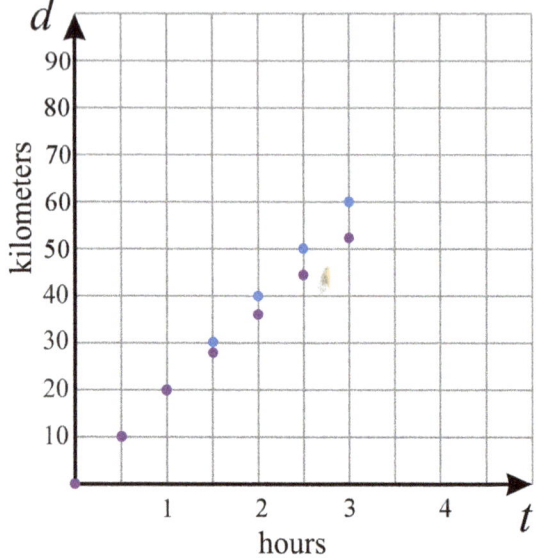

2.

a. $\dfrac{\$1.60}{\$3.00} = \dfrac{16}{30} = \dfrac{8}{15}$	b. $\dfrac{1.4 \text{ km}}{1.8 \text{ km}} = \dfrac{14}{18} = \dfrac{7}{9}$
c. $\dfrac{¼ \text{ mi}}{1 ½ \text{ mi}} = \dfrac{1 \text{ mile}}{6 \text{ miles}} = \dfrac{1}{6}$ (Multiply both terms of the initial ratio by 4.)	d. $\dfrac{⅔ \text{ L}}{1 \text{ L}} = \dfrac{2 \text{ L}}{3 \text{ L}} = \dfrac{2}{3}$ (Multiply both terms of the initial ratio by 3.)

3. a. −7.24   b. −0.3   c. −21.8

4.

a.	$6x + 3x + 1 = 9x − 2x − 7$ $9x + 1 = 7x − 7$ $2x + 1 = −7$ $2x = −8$ $x = −4$	b.	$16y − 4y − 3 = −4y − y$ $12y − 3 = −5y$ $17y − 3 = 0$ $17y = 3$ $y = 3/17$

Skills Review 42, p. 51

1.

a.	$-2x + 3x + 5 = 11x − 5x − 3$ $x + 5 = 6x − 3$ $−5x + 5 = −3$ $−5x = −8$ $x = 8/5 = 1\ 3/5$	b.	$40 + 20b − 4b = 2b + 50 − 7b$ $40 + 16b = 50 − 5b$ $40 + 21b = 50$ $21b = 10$ $b = 10/21$

Skills Review 42, cont.

2.

a. $-4 \div \dfrac{8}{9}$ $= -4 \cdot \dfrac{9}{8}$ $= \dfrac{-36}{8} = -4\dfrac{1}{2}$	b. $-\dfrac{11}{12} \div \left(-\dfrac{1}{3}\right)$ $= \dfrac{11}{12} \cdot \dfrac{3}{1}$ $= \dfrac{33}{12} = 2\dfrac{3}{4}$
c. $2.7 \div 0.004$ $= 2{,}700 \div 4$ $= 675$	d. $54.8 \div 0.11$ $= 5{,}480 \div 11$ ≈ 498.182

Skills Review 43, p. 52

1. a. The unit rate is that she earns $15/hour. She will earn $750 in 750/15 = <u>50 hours</u>.
   b. It will take her 17 days to earn at least $750.

2.

a.	$50x - 16 = -10 - 30x$ $80x - 16 = -10$ $80x = 6$ $x = 6/80 = 3/40$	b.	$-2k + 8 = 40 - 15k$ $13k + 8 = 40$ $13k = 32$ $k = 32/13 = 2\,6/13$

3.

a. $8 + (-2)$ $\downarrow$ $8 - 2 = 6$	b. $-5 - (-6)$ $\downarrow$ $-5 + 6 = 1$	c. $-10 + (-40)$ $\downarrow$ $-10 - 40 = -50$	d. $2 - (-9)$ $\downarrow$ $2 + 9 = 11$

4. a. $5x + y$ b. $12 - 8t$ c. $-19a - b$

5. a. Answers will vary; check the student's answer. For example, it is adding a row of two stars at the bottom, and adding one star to the top row, in each step.

   b. $5 + 3(n - 1)$, which simplifies to $3n + 2$

   c. $\quad 3n + 2 = 194$
      $\qquad\;\; 3n = 192$
      $\qquad\quad\; n = 64$

   There will be 194 snowflakes in <u>step 64</u>.

Skills Review 44, p. 53

1.

a.	$-3(x+7)$	$=$	$6x$	b.	$-8(y-4)$	$=$	$5(y+4)$
	$-3x-21$	$=$	$6x$		$-8y+32$	$=$	$5y+20$
	-21	$=$	$9x$		$-13y+32$	$=$	20
	$-21/9$	$=$	x		$-13y$	$=$	-12
	x	$=$	$-7/3 = -2\,1/3$		y	$=$	$12/13$

2. a. The expression $10a + 2b$ is for the perimeter, and $5a \cdot b$ for the area.

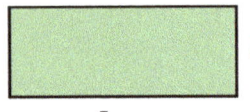 b

b. $5a$

3.

a. -21	b. -24	c. -28
45	-72	-20

4.

a. $\frac{7}{12}x = 42$	b. $\frac{6}{9}x = 7.2$
$7x = 504$	$6x = 64.8$
$x = 72$	$x = 10.8$

Skills Review 45, p. 54

1.

a. $-3 \div (-12)$	b. $18 \div (-24)$	c. $-16 \div (-20)$
$= \dfrac{-3}{-12} = \dfrac{3}{12} = \dfrac{1}{4}$	$= \dfrac{18}{-24} = -\dfrac{18}{24} = -\dfrac{3}{4}$	$= \dfrac{-16}{-20} = \dfrac{16}{20} = \dfrac{4}{5}$

2. a. The solution set is {11, 13, 15}.

   b. The solution set is {4, 6, 8, 10}.

3.

a.	$3x$	$=$	$-5 + 14 + (-3)$	b.	$6 \cdot (-8)$	$=$	$-4z$
	$3x$	$=$	6		-48	$=$	$-4z$
	x	$=$	2		48	$=$	$4z$
					z	$=$	12

4. a. $-|15| = -15$ b. $|-16| = 16$
   c. $-(6 + 8) = -14$ d. $-(7 - 4) = -3$

Skills Review 45, cont.

5. The student's proportion can vary from the answer given below, since it's possible to set up a proportion in several different correct ways. Check the student's answer.

a.
$$\frac{62 \text{ mi}}{x} = \frac{4.6 \text{ gal}}{9.4 \text{ gal}}$$
$$x \cdot 4.6 \text{ gal} = 62 \text{ mi} \cdot 9.4 \text{ gal}$$
$$x = \frac{62 \text{ mi} \cdot 9.4 \text{ gal}}{4.6 \text{ gal}}$$
$$x \approx 126.7 \text{ miles}$$

b.
$$\frac{3 \text{ lb}}{x} = \frac{1{,}000 \text{ ft}^2}{2{,}520 \text{ ft}^2}$$
$$x \cdot 1{,}000 \text{ ft}^2 = 3 \text{ lb} \cdot 2{,}520 \text{ ft}^2$$
$$x = \frac{3 \text{ lb} \cdot 2{,}520 \text{ ft}^2}{1{,}000 \text{ ft}^2}$$
$$x = 7.56 \text{ lb}$$

Skills Review 46, p. 55

1. a. $48 per coat   b. 20 cents per note card   c. $5.26 per pound

2.

a. $-4 \cdot (-6) = 24$	b. $7 \cdot (-3) = -21$	c. $-20 \cdot 9 = -180$
$24 \div (-6) = -4$	$-21 \div (-3) = 7$	$-180 \div 9 = -20$
or	or	or
$24 \div (-4) = -6$	$-21 \div 7 = -3$	$-180 \div (-20) = 9$

3. a. 28 min = 28/60 hr = 0.467 hr   b. 51 min = 51/60 hr = 0.85 hr

4. a. $1.6 \cdot 10^4$ b. $3.07 \cdot 10^5$
   c. $4.29 \cdot 10^4$ d. $9.108 \cdot 10^6$

5.

a. $\dfrac{9}{-2\frac{1}{4}}$	b. $\dfrac{-\frac{4}{5}}{6.4}$	c. $\dfrac{3.25}{\frac{2}{5}}$
$= 9 \div \left(-\frac{9}{4}\right) = 9 \cdot \left(-\frac{4}{9}\right)$	$= -0.8 \div 6.4 = -8 \div 64$	$= 3.25 \div 0.4 = 32.5 \div 4.$
$= -4$	$= -1/8 = -0.125.$	Now use long division to get the final answer of 8.125.
	Or, you can use long division to divide $-8 \div 64 = -0.125.$	

6. a. −3   b. 1

Skills Review 47, p. 56

1. Let p be the price of each individual painting. Inequality: $12p - 45 \geq 400$.
   Solution: $12p \geq 445$ from which $p \geq 37.08\overline{3}$. Each painting has to cost at least $37.09.

2. a. $cost = 25n + 12m$ b. $price = (4/5) \cdot \$30$ or $0.8 \cdot \$30$. There are other possibilities also.

3. $y = x - 2$

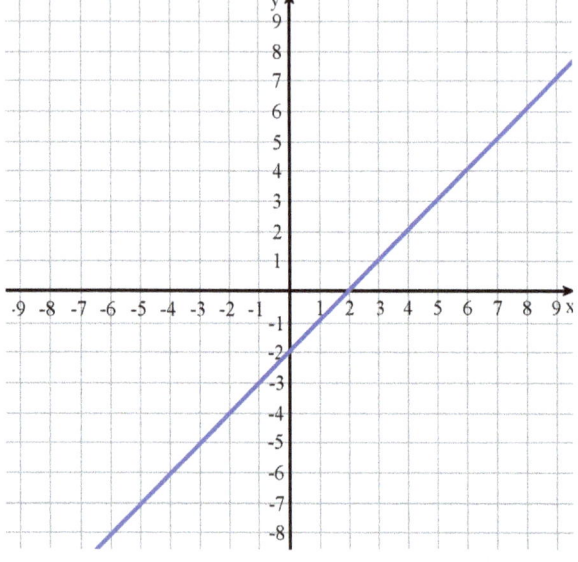

x	−4	−3	−2	−1	0	1	2	3
y	−6	−5	−4	−3	−2	−1	0	1

The variables are not in direct variation.
We can see that, for example, from the fact that
the point (0, 0) is not included in the table.

4. a. $x = -25.56$ b. $s = 0.63$

Skills Review 48, p. 57

1. a. Check by substituting the values in the equation:

   $y = 3x - 9$

   $2 = 3(-3) - 9$

   $2 = 0$

   This is a false equation, so the number pair does not fulfill the equation.

   b. Check by substituting the values in the equation:

   $y - 6 = -x$

   $-2 - 6 = -5$

   $-8 = -5$

   This is a false equation, so the number pair
   does not fulfill the equation.

2. a. Each time the x-values increase by one, the y-values
   increase by ½, so the slope is ½.

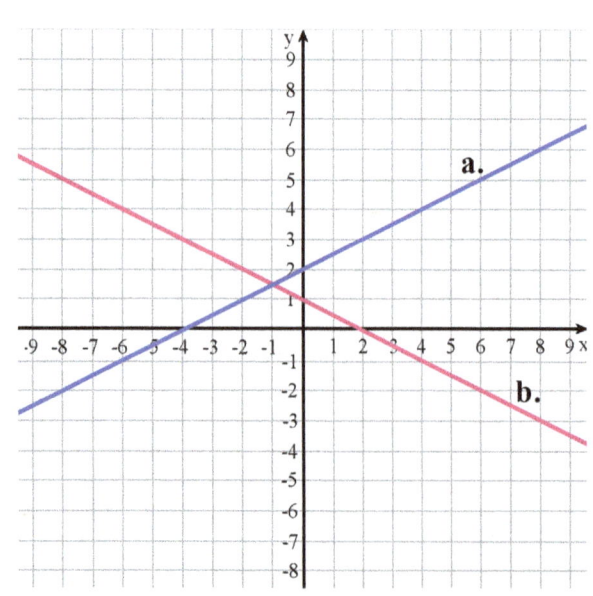

x	−5	−4	−3	−2	−1	0	1
y	−½	0	½	1	1½	2	2½

   b. Each time the x-values increase by one, the y-values
   decrease by ½, so the slope is −½.

x	−6	−4	−2	0	2	4	6
y	4	3	2	1	0	−1	−2

Skills Review 48, cont.

3. The side of the square is 8 cm. When it is shrunk, its side becomes 6 cm. Now the area of the square is 36 cm².

4.

a.	$\dfrac{x+7}{4}$	=	18	b.	$\dfrac{x-3}{8}$	=	−5
	$x+7$	=	72		$x-3$	=	−40
	x	=	65		x	=	−37

Chapter 7: Percent

Skills Review 49, p. 58

1. a. −7   b. 44   c. −29

2. a.

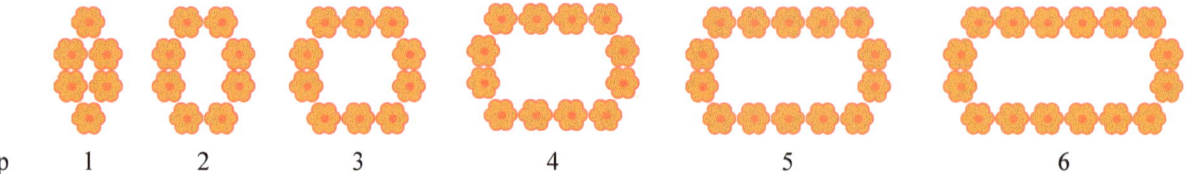

Step   1   2   3   4   5   6

b. Answers will vary. Check the student's answer. For example: It adds one flower to the top and to the bottom side.
c. 82 flowers
d. Step 1 has 6 flowers, and after that it adds 2 each time. So the formula is $6 + 2(n − 1)$, which simplifies to $2n + 4$.
   Or, you can see the pattern as having exactly n flowers both in its top and bottom rows, plus four additional flowers.

3. a. $d = 90t$

b.
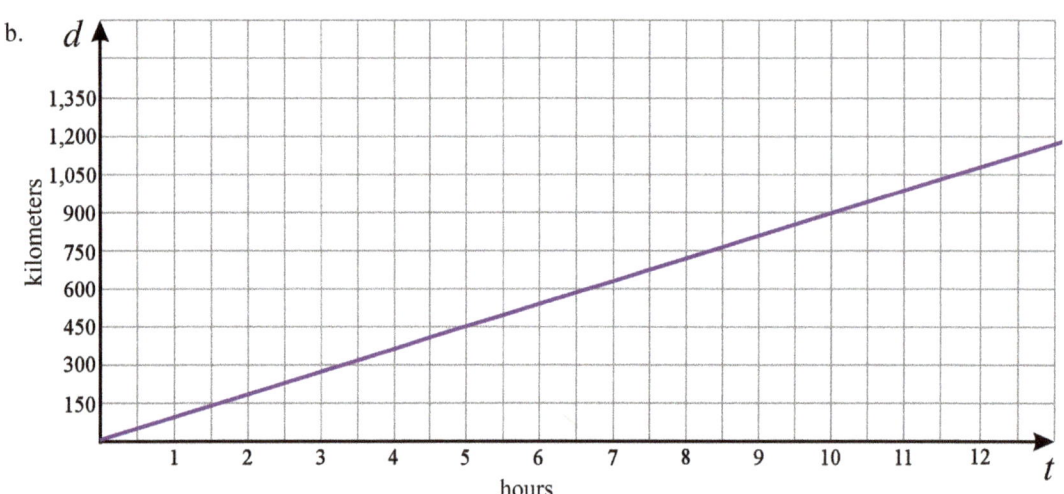

c. Three hours 15 minutes is 3.25 hours. We use the equation $d = 90t$ to get $d = 90$ km/h × 3.25 h = <u>292.5 km</u>.

d. The bus will take 1,270 / 90 h = 14.111... hours ≈ 14 hours 7 minutes for the trip. Leaving Chicago at 6:45 am, it would arrive at 8:52 PM. (Of course, in real life the bus would need to stop in between, thus taking longer.)

Skills Review 50, p. 59

1. a. In real life, the room measures 10 ft by 12 ft. In scale 1 in : 3 ft, it would measure 10/3 in by 4 in, or 3 1/3 in by 4 in.
(To actually draw 3 1/3 inches, depending on the tools at hand, you may have to approximate it as 3 5/16 inches.)

   b. In scale 1 in : 5 ft, it would measure 10/5 in by 12/5 in, or 2 in by 2 2/5 in.
(To actually draw 2 2/5 inches, depending on the tools at hand, you may have to approximate it as 2 3/8 inches.)

2.

a.	$2x - 76 = -4 - 7x$	b.	$-4y - 9 = 63 + 2y$
	$9x - 76 = -4$		$-9 = 63 + 6y$
	$9x = 72$		$-72 = 6y$
	$x = 8$		$-12 = y$
			$y = -12$

3. a. 60 km per hour   b. 13 inches : 5 minutes

4.

a. $-7 + (-10) = -17$	b. $12 - (-9) = 21$	c. $2 + (-6) = -4$	d. $5 + 8 = 13$
↓	↓	↓	↓
$-7 - 10 = -17$	$12 + 9 = 21$	$2 - 6 = -4$	$5 - (-8) = 13$

5. a. They are equal.   b. $0.\overline{75}$ is more. It is $0.00\overline{75}$ more than 0.75.

Skills Review 51, p. 60

1.

a. $39.4\% = \dfrac{394}{1000} = 0.394$	b. $21.6\% = \dfrac{216}{1{,}000} = 0.216$	c. $70.2\% = \dfrac{702}{1000} = 0.702$

2.

on map (cm)	in reality (cm)	in reality (m)	in reality (km)
1 cm	20,000	200	0.2
4 cm	80,000	800	0.8
6.5 cm	130,000	1300	1.3
0.9 cm	18,000	180	0.18
12.7 cm	254,000	2540	2.54

3.

a. $-6 \cdot 8 = -48$	b. $(-2) \cdot (-9) = 18$	c. $(-10) \cdot 30 \cdot (-2) = 600$
$-4 \cdot (-7) = 28$	$5 \cdot (-12) = -60$	$-3 \cdot (-50) \cdot 4 = 600$

4. a. $9x^2y^4$ b. $3a + 3d$ c. $3z + 5$

5.

a.	$5 - (-12) = x + 9$	b.	$3 - 10 = 4 + w$
	$17 = x + 9$		$-7 = 4 + w$
	$8 = x$		$-11 = w$
	$x = 8$		$w = -11$

Skills Review 52, p. 61

1.

a.			b.			
	$4(x+3)$	$= 3(x-(-1))$		$6(y-9)+3$	$=$	15
	$4x+12$	$= 3(x+1)$		$6y-54+3$	$=$	15
	$4x+12$	$= 3x+3$		$6y-51$	$=$	15
	$x+12$	$= 3$		$6y$	$=$	66
	x	$= -9$		y	$=$	11

2. a. $15s$ b. $14p^2$ c. $7x^2$

3. Let x be the area of the original land. Equation: $x/5 = 1{,}800$ ha. Solution: $x = 9{,}000$ ha.

4. Answers will vary. Check the student's answer. For example:

   a. The price must be set to be more than $85.

   b. We can host at most 40 people at the party.

5. In 60 hours. For example, you can think this way: Divide the $120/9 hr by six to get $20 per 1 ½ hr. Then multiply that by four, and then by 10, to get the rate of $800 per 60 hours.

Earnings	20	80	120	800
Work Hours	1 ½	6	9	60

Skills Review 53, p. 62

1.

a.
$$\frac{T}{30} = \frac{72}{8}$$
$$8T = 2160$$
$$T = 270$$

b.
$$\frac{13}{152} = \frac{4}{M}$$
$$13M = 608$$
$$M \approx 46.77$$

2. a. New price = $1.08 \cdot \$594.85 = \$642.44$

   b. New price = $1.045 \cdot \$769.90 = \$804.55$

3.

a. The unit rate can be calculated either for so many feet per 1 dollar, or so many dollars per 1 foot.

Feet per dollar: $\dfrac{5\,¼\text{ ft}}{\$3} = \dfrac{21/4\text{ ft}}{\$3} = \dfrac{21}{4} \cdot \dfrac{1}{3}$ ft/dollar $= \dfrac{7}{4}$ ft/dollar $= 1\dfrac{3}{4}$ ft/dollar.

Dollars per foot: $\dfrac{\$3}{5\,¼\text{ ft}} = \dfrac{\$3}{21/4\text{ ft}} = \dfrac{3}{1} \cdot \dfrac{4}{21}$ dollars/ft $= \dfrac{4}{7}$ dollars/ft or about $0.57 per foot.

b. The unit rate can be calculated either for so many hours per 1 documentary, or so many documentaries per 1 hour.

Hours per documentary: $\dfrac{3/4\text{ hr}}{2/3\text{ docum.}} = \dfrac{3}{4} \cdot \dfrac{3}{2}$ hr/docum. $= \dfrac{9}{8}$ hr/documentary $\approx$ 1 hour 8 min/documentary

Documentaries per hour: $\dfrac{2/3\text{ docum.}}{3/4\text{ hr}} = \dfrac{2}{3} \cdot \dfrac{4}{3}$ docum/hr $= \dfrac{8}{9}$ docum/hour $\approx$ 0.89 documentaries per hour

Skills Review 54, p. 63

1.

a. $d = v\ t$	b. $d = v\ t$
$\quad\ \downarrow\ \ \downarrow\ \downarrow$	$\quad\ \downarrow\ \ \downarrow\ \downarrow$
$215 = 95\ t$	$35 = 42\ t$
$t = \dfrac{215}{95}$ hr ≈ 2.2632 hr	$t = \dfrac{35}{42}$ hr $= \dfrac{5}{6}$ hr $= 50$ min
$\approx\ 2$ hr 16 min	Beth will arrive at 7:55 am.

2. a. They are not in proportion. As the number of people increases, the time *decreases*.

3. a. $349.80/p = 66/100$

b. $\dfrac{349.80}{p} = \dfrac{66}{100}$

$\quad 66p = 34{,}980$

$\quad\quad p = 530$

The price before the discount was $530.

4. a. −0.66 b. −6.91 c. 10.4

Skills Review 55, p. 64

1. a. $2.45 \cdot 10^8$ b. $6.27 \cdot 10^3$

2.

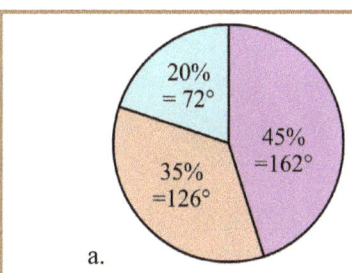
a.

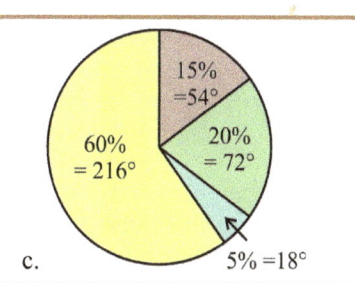
b.

c.

Pie chart a: 20% = 72°, 45% = 162°, 35% = 126°

Pie chart b: 20% = 72°, 1/8 = 45°, 55% = 198°, 1/8 = 45°

Pie chart c: 15% = 54°, 20% = 72°, 5% = 18°, 60% = 216°

3. Answers will vary. For example:

a. $2 + (-9) + 2 = -5$	b. $-10 + 6 + 5 = 1$
$\ \ -11 + 4 + 2 = -5$	$\ \ 2 + 1 + (-2) = 1$

4.

a. $x + \dfrac{1}{4} = \dfrac{11}{20}$	b. $x - \dfrac{2}{9} = \dfrac{1}{6}$
$\quad x = \dfrac{11}{20} - \dfrac{1}{4} = \dfrac{3}{10}$	$\quad x = \dfrac{1}{6} + \dfrac{2}{9} = \dfrac{7}{18}$

5. a. 1/3 b. −1 1/2 c. 4/7

Skills Review 56, p. 65

1. Let n be the number of blouses you buy. Then $13.45n - 15$ is the total cost with the discount.
   The inequality is: $13.45n - 15 \leq 80$. From that we get $13.45n \leq 95$, and $n \leq 7.063...$.
   You can buy at most 7 blouses.

2. a. apples = $x/3$ b. bananas = $4x$

3. The first shampoo costs $0.32/fl. oz and the second $0.31/fl. oz. The first shampoo costs more per fluid ounce.

4.
> $15 \div 3 = 5$
> $9 \div 3 = 3$
> $3 \div 3 = 1$
> $(-3) \div 3 = -1$
> $(-9) \div 3 = -3$
> $(-15) \div 3 = -5$
> $(-21) \div 3 = -7$
> $(-27) \div 3 = -9$

5. The discount percentage is $\dfrac{\text{difference}}{\text{original}} = \dfrac{350 - 308}{350} = \dfrac{42}{350} = 0.12 = \underline{12\%}$.

Chapter 8: Geometry

Skills Review 57, p. 66

1. Answers will vary. Check the student's answers. All such lines are parallel, and obey the rule that for every 5 horizontal units, they will rise by 3 vertical units. Some examples:

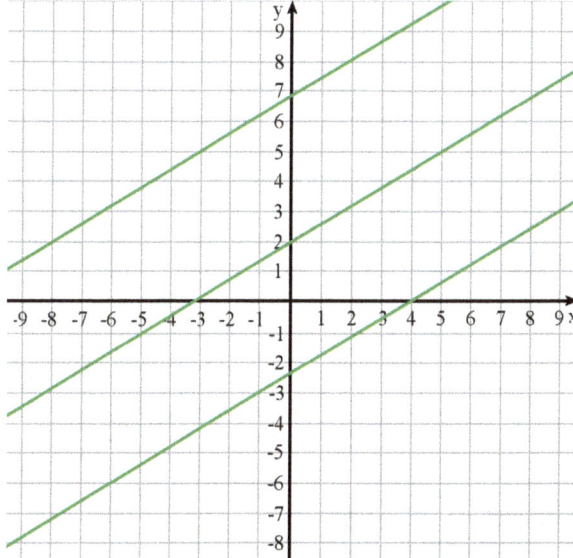

2. Answers will vary. Check the student's answers. For example:

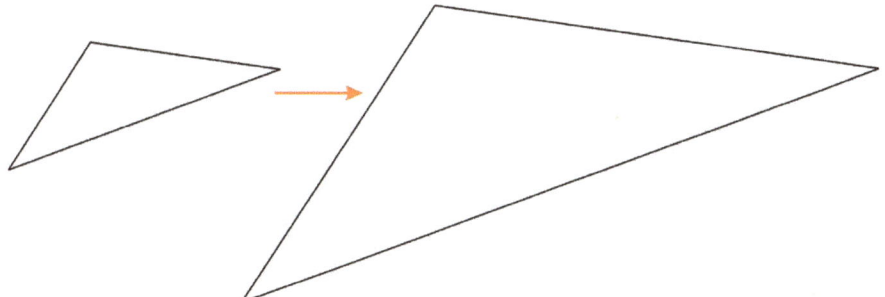

3.

a. $-\dfrac{1}{7} \cdot 0.8$	b. $\dfrac{1}{8} \cdot (-2.6)$	c. $0.4 \cdot \dfrac{3}{5}$
$= -\dfrac{1}{7} \cdot \dfrac{8}{10} = -\dfrac{8}{70} = -\dfrac{4}{35}$	$= -2.6 \div 8$	$= \dfrac{2}{5} \cdot \dfrac{3}{5} = \dfrac{6}{25}$
or $= -0.8 \div 7 \approx -0.114286$	$= -0.325$	or $= 0.4 \cdot 0.6 = 0.24$

4.

Four-ninths of a number is 5.62. What is the number?	
Equation:	Another way:
$\dfrac{4}{9}x = 5.62$ $4x = 50.58$ $x = 12.645$	Since 4/9 of this number is 5.62, then one-ninth of it is $5.62 \div 4 = 1.405$. And then, the number itself is $9 \cdot 1.405 = 12.645$.

Skills Review 58, p. 67

1. a.

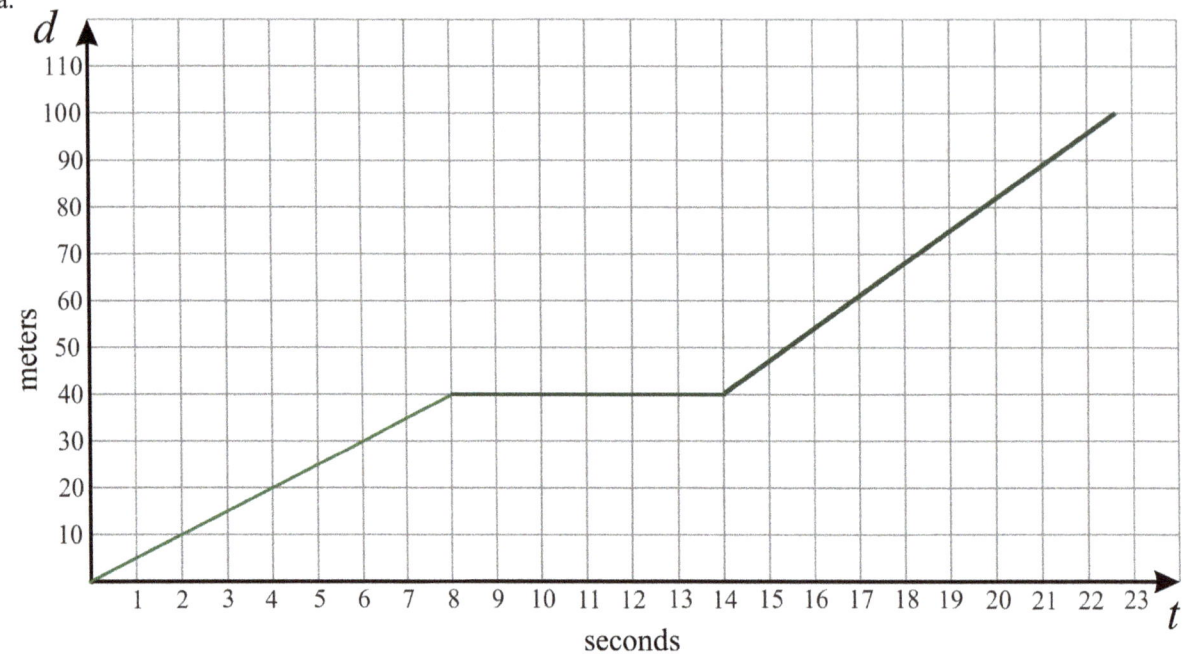

b. He first runs for eight seconds, and advances 5 m/s · 8 s = 40 meters before falling. Then he has 60 meters left to run, at a speed of 7 m/s. This takes him 60 m ÷ 7 m/s ≈ 8.6 seconds. So his total time is 8 s + 6 s + 8.6 s = 22.6 seconds.

2. a. −0.082   b. 1.986   c. 0.516

3. Their average area is (96,000 + 69,000)/2 = 82,500 square miles.
   The relative percentage difference between their areas is (96,000 − 69,000)/82,500 = $0.32\overline{72}$ ≈ 32.7%.

4.

a.	$-8 = \dfrac{-6x}{9}$	b.	$3 - 5 = \dfrac{s+7}{4}$
	$-72 = -6x$		$-2 = \dfrac{s+7}{4}$
	$12 = x$		$-8 = s+7$
	$x = 12$		$-15 = s$
			$s = -15$

Skills Review 59, p. 68

1.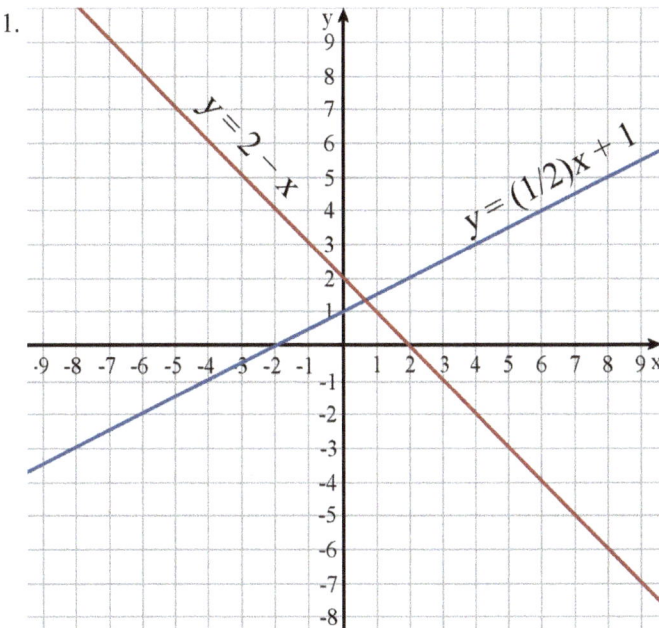

2. The width of the room on paper is 2.8 m · (4 cm/1 m) = 11.2 cm.
   The length of the room on paper is 4 · (4 cm/1 m) = 16 cm.

   The depth of the desk on paper is 0.6 m · (4 cm/1 m) = 2.4 cm.
   The width of the desk on paper is 0.9 m · (4 cm/1 m) = 3.6 cm.

   The image on the right is not to scale. Please check the student's work.

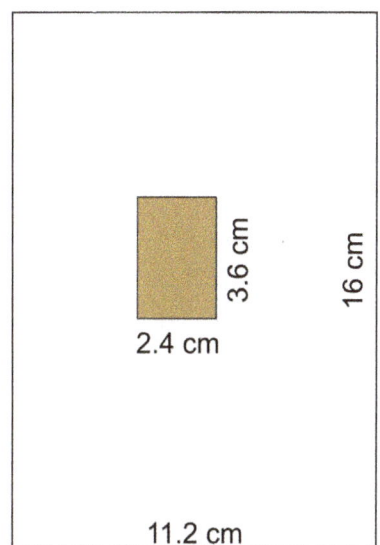

3.
Angle	Degrees	Fraction	Percentage
α	72°	1/5	20%
β	150°	5/12	41.7%
γ	58°	29/180	16.1%
δ	80°	2/9	22.2%

4.
a. 6 to 42 = 1 to _7_	b. 4 : _32_ = 12 : 96	c. 160 : 400 = 2 : _5_	d. $\dfrac{7}{11} = \dfrac{91}{143}$

5.
a. Camera: $70; 8% sales tax. Tax to add: $ _5.60_ Price after tax: $ _75.60_	b. Sofa: $300; 5% sales tax. Tax to add: $ _15_ Price after tax: $ _315_	c. Game: $40; 7% sales tax. Tax to add: $ _2.80_ Price after tax: $ _42.80_

Skills Review 60, p. 69

1. a. $\dfrac{7}{12} = 0.58\overline{3}$ b. $\dfrac{-3}{8} = -0.375$ c. $\dfrac{65}{-100} = -0.65$

2. a. $|-183| = 183$ b. $|x|$

3. The town is 3.2 in · 300,000 = 960,000 inches long. And 960,000 in = 80,000 ft ≈ <u>15 miles</u>.

4.

a.	$4x + 5x + 2$	$=$	$12x - 2x - 7$	b.	$15y - 9y - 6$	$=$	$9y - y$
	$9x + 2$	$=$	$10x - 7$		$6y - 6$	$=$	$8y$
	2	$=$	$x - 7$		$-2y - 6$	$=$	0
	x	$=$	9		$-2y$	$=$	6
					y	$=$	-3

5. a. Interest: <u>$8,000 · 0.07 · 1 = $560</u>   Total to withdraw: <u>$8,650</u>

   b. Interest: <u>$6,300 · 0.054 · 3 = $1020.60</u>   Total to withdraw: <u>$7,320.60</u>

Skills Review 61, p. 70

1. a. New price = 0.68 · $45   b. New price = $0.83p$

2. a. $-0.7w - 12.6$ b. $-1.8x + 2.6$ c. $-200x - 320$

3. a. The ratio of doctors : people is 75,000 : 205,700,000. To get the ratio of doctors per 10,000 people, we note that 205,700,000 is 20,570 times 10,000, so the ratio of doctors per 10,000 people simply is 75:000 : 20,570. Now we write the latter as a unite rate by dividing, and get 3.6461 doctors per 10,000 people, or about <u>3.6 doctors per 10,000 people</u>.

   b. We expect to find 6.8 · 430,000/10,000 ≈ <u>292 doctors</u> in that area.

4. The other bottom angle of the triangle is 180° − 144° = 36°. Then, the top angle of the triangle is 180° − 65° − 36° = 79°.

5. a. −12   b. −53   c. −46

6.

Variety	Amount sold	Percentage of total	Central Angle
fava	16	15.4	55
kidney	29	27.9	100
navy	22	21.2	76
pinto	37	35.6	128
TOTALS	104	100.1%	359°

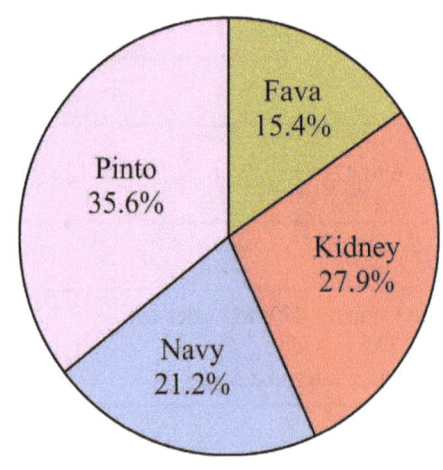

Skills Review 62, p. 71

1.

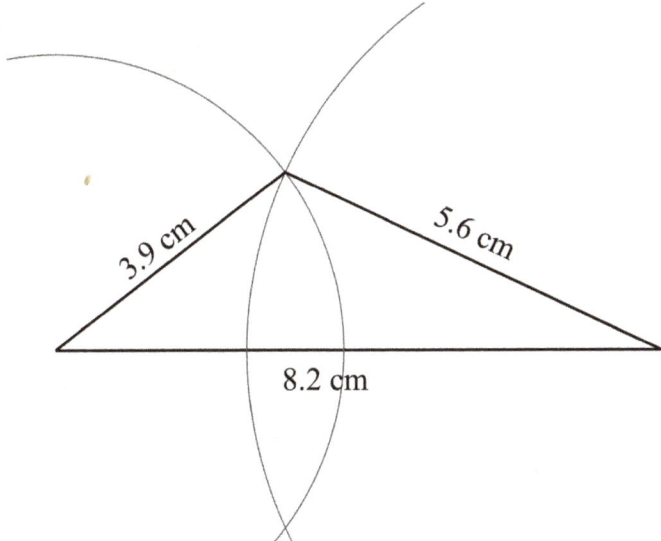

2. Answers will vary. Check the student's answer. For example:

   a. A fish was at a depth of −40 ft, and then it descended another 10 feet deeper into the lake.

   b. Mark had $30 in his piggy bank. Then he broke something that cost $55 to fix.

3.

a.	$-6 + r = -4 + (-9)$	b.	$12 - (-5) = 6 + 8 + t$
	$-6 + r = -13$		$17 = t + 14$
	$r = -7$		$t = 3$

4. $-e$ or $-1e$

5.

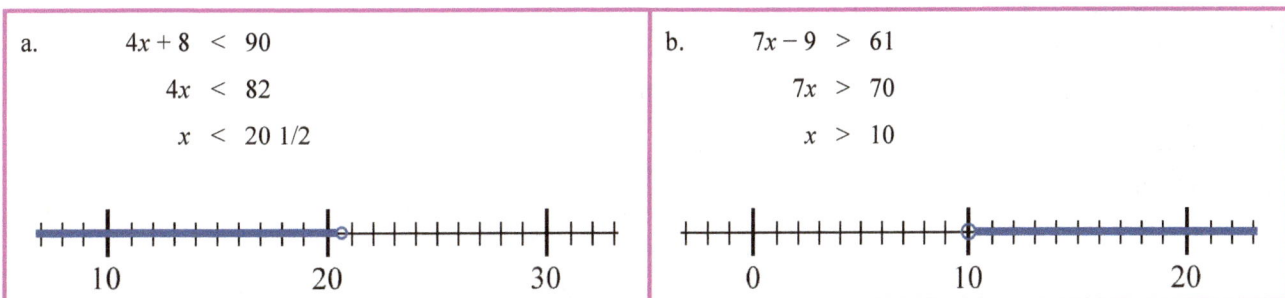

a. $4x + 8 < 90$
    $4x < 82$
    $x < 20\ 1/2$

b. $7x - 9 > 61$
    $7x > 70$
    $x > 10$

Skills Review 63, p. 72

1. a. The new price is $0.8 \cdot \$50 = \$40$.
   b. The discount amount is $15. Since $5 is 10% of $50, then $15 is 30% of $50. So the discount is 30%.

2. a. $720/2000 = 72/200 = 36/100 = 9/25$
   b. $(9/25) \cdot 240 = 86.4$. So we can expect about 86 people to like roses.

3. It helps to first add all the positives and all the negatives separately, then combine those.
   a. −5.6   b. −$6.88

Skills Review 63, cont.

4. a. No. There are multitudes of such triangles. Though they are all similar to each other (with the same basic shape), they are not necessarily congruent since they can be of different sizes.

   b. Yes.

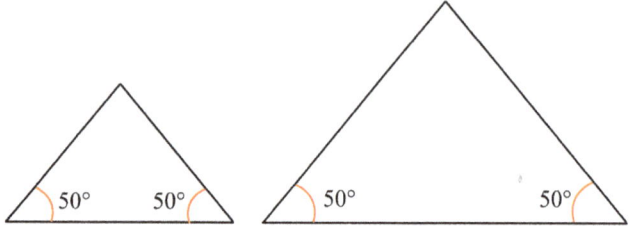

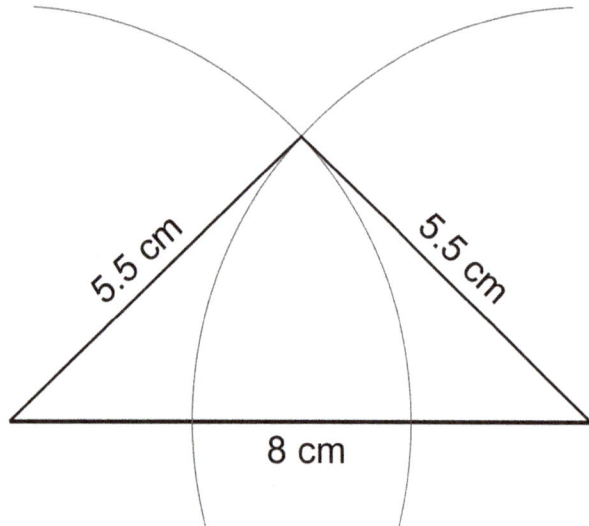

Skills Review 64, p. 73

1. a. −8   b. −15   c. −20

2. Barry is $\frac{55}{135}$ = 41% taller than Seth. Seth is $\frac{55}{190}$ = 29% shorter than Barry.

3. a. Circumference = (22/7) · 7 cm = 22 cm
   b. Circumference = (22/7) · 10 in = 31.4 in
   c. Circumference = (22/7) · 3 m = 9.4 m

4. The way the student's proportion is set up may differ from the solutions below, but the final answer should be the same.

   a.
   $$\frac{\$1330}{140 \text{ ft}^2} = \frac{x}{260 \text{ ft}^2}$$
   $$140 \text{ ft}^2 \cdot x = \$1330 \cdot 260 \text{ ft}^2$$
   $$x = \frac{\$1330 \cdot 260 \text{ ft}^2}{140 \text{ ft}^2}$$
   $$x = \$2{,}470$$

   It costs $2,470 to pave 260 square feet of driveway.

   b.
   $$\frac{74 \text{ mi}}{4.8 \text{ gal}} = \frac{x}{6.5 \text{ gal}}$$
   $$4.8 \text{ gal} \cdot x = 74 \text{ mi} \cdot 6.5 \text{ gal}$$
   $$x = \frac{74 \text{ mi} \cdot 6.5 \text{ gal}}{4.8 \text{ gal}}$$
   $$x = 100.208\overline{3} \text{ mi}$$

   It can go 100.2 miles on 6.5 gallons of gasoline.

5. a. 2 hours 38 minutes   b. 27 minutes

Skills Review 65, p. 74

1. a. $x = 35$ b. $x = -288$ c. $x = -55$

2. a. The final amount on her account is the principal plus interest. We can write this as

$$\$15{,}400 = p + I$$
$$\$15{,}400 = \$8{,}800 + prt$$
$$\$15{,}400 = \$8{,}800 + \$8{,}800 \cdot 0.075 \cdot t$$
$$\$6{,}600 = \$8{,}800 \cdot 0.075 \cdot t$$
$$\$8{,}800 \cdot 0.075 \cdot t = \$6{,}600$$
$$t = \frac{\$6{,}600}{\$8{,}800 \cdot 0.075}$$
$$t = 10$$

In <u>10 years</u>, her account contained $15,400.

b. Since the interest amount was $400, and the time was two years, in one year the account accumulated $200. And $200 is $200/8000 = 2/80 = 1/40 = 0.025$, or <u>2.5%</u>.

You can also use the formula I = *prt*:

$$I = prt$$
$$\$400 = \$8{,}000 \cdot r \cdot 2$$
$$\$400 = \$16{,}000 r$$
$$r = \frac{\$400}{\$16{,}000}$$
$$r = 0.025$$

3.

x	−6	−5	−4	−3	−2	−1	0	1
y	9	7½	6	4½	3	1½	0	−1½

In proportion or not? <u>Yes.</u>

Unit rate: <u>−1 ½ or −3/2</u>

Equation: <u>y = (−3/2)x</u>

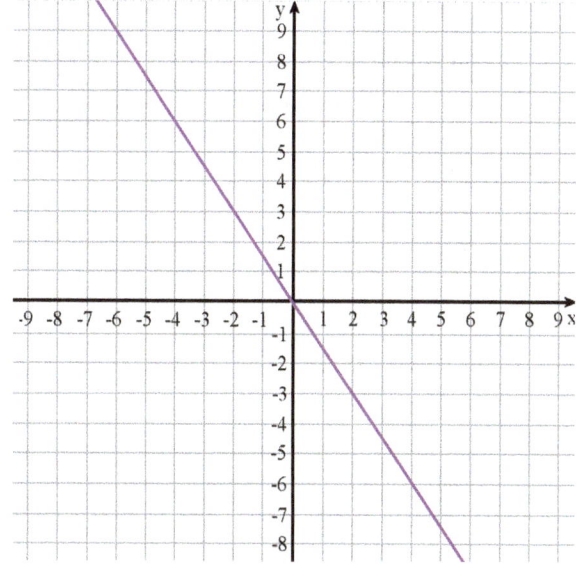

4. The expression (b) or the expression (d).

Skills Review 66, p. 75

1. a. −4/3   b. 7/4.5 = 14/9

2. Jeremy drives for 10 h − 1 h 15 min = 8 h 45 min, or 8.75 hr, each day. Let v be the unknown speed. Since the distance (d) has to be at least 500 miles, and since distance is given with the formula $d = vt$, we can write the basic inequality $vt \geq 500$. Substituting 8.75 hr for the time, we get: $8.75v \geq 500$.
From this, $v \geq 500/8.75 \approx 57.14$ mph.

He has to drive, on average, at the speed of about 58 miles per hour, in order to travel at least 500 miles per day.

3.

Expression	the terms in it	coefficient(s)	Constants
$(3/7)s$	$(3/7)s$	3/7	none
$0.5x + 9.2y$	$0.5x$ and $9.2y$	0.5 and 9.2	none
$x^3y^8 + 6$	x^3y^8 and 6	1	6

4. The area of a trapezoid is given with $A = (a + b)/2 \cdot h$. Using that, we can write the equation

$$4125 \text{ m}^2 = \frac{70 \text{ m} + 95 \text{ m}}{2} \cdot h$$

$$4125 \text{ m}^2 = 82.5 \text{ m} \cdot h$$

$$h = \frac{4125 \text{ m}^2}{82.5 \text{ m}} = 50 \text{ m}$$

5. a. −7/9   b. 6 4/15   c. −2

Skills Review 67, p. 76

1. a. $-11 + 7 + 3 = -1$ b. $8 + 4 - 9 = 3$

2. a. From $d = vt$, we can solve that $t = d/v$. The first part of the trip took him 560 km / (250 km/h) = 2.24 hours or about 2 hours 14 minutes.

   b. The second part took him 320 km / (250 km/h) = 1.28 hours or about 1 hour 17 minutes.

   c. He reached Tampa at 9:15 + 2 hours 14 minutes + 2 hours + 1 hour 17 minutes = 14:46 or 2:46 PM.

   d. Notice: his total trip time was very close to 5 1/2 hours.

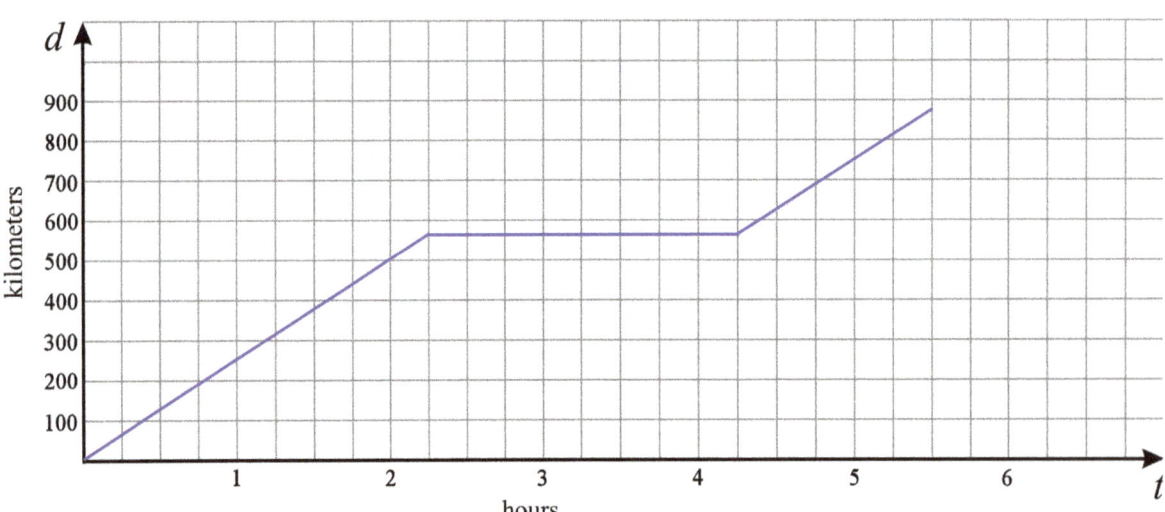

Skills Review 67, cont.

3.

a. $\dfrac{14}{-2\frac{1}{4}}$	b. $\dfrac{-\frac{5}{8}}{3.6}$	c. $\dfrac{5.75}{\frac{2}{10}}$
$= 14 \div \left(-\dfrac{9}{4}\right)$	$= -\dfrac{5}{8} \div 3.6 = -\dfrac{5}{8} \div \dfrac{36}{10}$	$= 5.75 \div \dfrac{1}{5} = 5.75 \cdot 5$
$= \dfrac{14}{1} \cdot \left(-\dfrac{4}{9}\right) = -\dfrac{56}{9} = -6\dfrac{2}{9}$	$= -\dfrac{5}{8} \cdot \dfrac{10}{36} = \dfrac{5}{4} \cdot \dfrac{5}{36} = \dfrac{25}{144}$	$= 28.75$
With a calculator, do $14 \div -2.25 = -6.222...$	With a calculator, do $-5 \div 8 \div 3.6 = -0.1736111...$	With a calculator, do $5.75 \cdot 10 \div 2 = 28.75$.

4. a. Check the student's cone.

b. The surface area of the cone is the same as the area of the half circle. $A = \frac{1}{2} \cdot \pi (9 \text{ cm})^2 \approx 127 \text{ cm}^2$.

Skills Review 68, p. 78

1.

a. $-9 \cdot (-4) = \underline{36}$	b. $3 \cdot (-12) = -36$	c. $-60 \cdot 7 = -420$
$36 \div (-9) = -4$ or $36 \div (-4) = -9$	$-36 \div 3 = -12$ or $-36 \div (-12) = 3$	$-420 \div 7 = -60$ or $-420 \div (-60) = 7$

2. a. Yes, it does: $-20 = 3(-6) - 2$ is a true equation.

b. No, it isn't: $-6 - 2 = -4$ is not a true equation.

3. a. The dimensions of the bathroom in the plan are 3.7 m · (4 cm/1 m) = <u>14.8 cm</u> and 4.6 m · (4 cm/1 m) = <u>18.4 cm</u>.

b. The dimensions of the living room in reality are 26.8 cm · (1 m/ 4 cm) = <u>6.7 m</u> and 31.6 cm · (1 m/ 4 cm) = <u>7.9 m</u>.

4.

a. A rectangle with 7.0 ft and 3.8 ft sides → A = <u>26.6</u> sq ft

Conversion: A = <u>26.6</u> sq ft · $\dfrac{144 \text{ sq in}}{1 \text{ sq ft}}$ = <u>3,830.4</u> sq in.

b. A rectangle with 18 ft and 12 ft sides → A = <u>216</u> sq ft

Conversion: A = <u>216</u> sq ft · $\dfrac{1 \text{ sq yd}}{9 \text{ sq ft}}$ = <u>24</u> sq yd.

5.

a. $\dfrac{200 \text{ cm}}{4.7 \text{ m}} = \dfrac{200 \text{ cm}}{470 \text{ cm}} = \dfrac{20}{47}$	b. $\dfrac{5 \text{ ft } 9 \text{ in}}{2 \text{ ft } 6 \text{ in}} = \dfrac{69 \text{ in}}{30 \text{ in}} = \dfrac{23}{10}$

Skills Review 69, p. 79

1. ∠α = 20°   ∠β = 160°   ∠γ = 71°

2. a. $c = -3725.82$ b. $x = 2.518$

3. In square kilometers, the area is 2.7 km · 1.9 km = 5.13 km². In hectares, this is 513 ha.

4. She will take 3 hours 45 minutes to ride 27 miles.

$$\frac{6 \text{ miles}}{50 \text{ minutes}} = \frac{3 \text{ miles}}{25 \text{ minutes}} = \frac{27 \text{ miles}}{225 \text{ minutes}}$$

5. a. $-0.7w - 16.8$ b. $4x - 30$ c. $-60x - 270$

6.

Scientific Notation	(in-between calculation)	Common notation
$6.45 \cdot 10^9$	6.45 · 1,000,000,000	6,450,000,000
$4.212 \cdot 10^6$	4.212 · 1,000,000	4,212,000
$7.3936 \cdot 10^2$	7.3936 · 100	739.36

Skills Review 70, p. 80

1. The exact answers are not fully exact in (a) and (c), as they are rounded to three decimals, as asked in the problem.

a. $\frac{3}{15} \cdot (-2.074)$	b. $8.62 \cdot \left(-1\frac{5}{9}\right)$	c. 22% of $4\frac{3}{8}$
Estimate: 1/5 of −2 = −0.4	Estimate: 9 · (−1.5) = −13.5	Estimate: 1/5 of 4.5 = 0.9
Exact: −2.074/5 = −0.415	Exact: 8.62 · (−14/9) = −13.40$\overline{8}$	Exact: 0.22 · 35/8 = 0.963

2. a. $2 - 6x$ or $-6x + 2$ b. $11 - 3b$ or $-3b + 11$
   c. $2 - 3z$ or $-3z + 2$ d. $5f - 20$ or $-20 + 5f$

3. a. $\dfrac{4\frac{1}{2} \text{ C oats}}{5 \text{ doz cookies}} = \dfrac{4\frac{1}{2} \text{ C oats}}{60 \text{ cookies}} = \dfrac{9}{2} \cdot \dfrac{1}{60}$ cups of oats/cookie $= \dfrac{3}{40}$ cups of oats/cookie

   b. $30.10 ÷ 3.5 lb = $8.60 per pound

4. The side of the square is 9 cm. When the square is shrunk, its sides become 6 cm. The area is now 36 cm².

5.

a. $\frac{3}{5}x = 480$	b. $\frac{5}{8}y = \$26.74$
$3x = 2{,}400$	$5y = \$213.92$
$x = 800$	$y = \$42.784$

Chapter 9: Pythagorean Theorem

Skills Review 71, p. 81

1. On a map with a scale of 1:350,000 the distance would be
   67.39 mi ÷ 350,000 · 5,280 ft/mi · 12 in/ft ≈ 12.2 inches.

   On a map with a scale of 1:400,000 the distance would be
   67.39 mi ÷ 400,000 · 5,280 ft/mi · 12 in/ft ≈ 10.7 inches.

2.

a. $\dfrac{x}{4} - (-14) = -9 \cdot 6$ $\dfrac{x}{4} + 14 = -54$ $\dfrac{x}{4} = -68$ $x = -272$	b. $5 - 9p = -1.3$ $-9p = -6.3$ $p = 0.7$

3.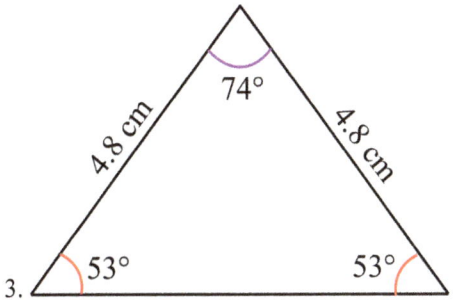

   The base angles measure 53°.

4. a. The cross section is a circle. b. The cross section is a rectangle. c. The cross section is a triangle.

Skills Review 72, p. 82

1. a. The final price is $35.99 · 0.78 · 1.08 = $30.32.

   b. The final price is $157.95 · 0.85 · 1.06 = $142.31.

2. a. The set 11 cm, 4 cm, 6 cm won't make a triangle.

   b. Answers will vary; check the student's answer. Check that any two side lengths added together is more than the third side. For example, if we change the 4-cm side to 7 cm, it works, because 11 cm + 7 cm > 6 cm, and 11 cm + 6 cm > 7 cm, and 6 cm + 7 cm > 11 cm.

Skills Review 72, cont.

3. a. Let P symbolize his pay and *t* his work time in hours. Then $P = 28t$.

   b. and c. Answers will vary; check the student's answer. Check that the point (15, 240) fits on the grid. Check that the point (1, 28) is marked.

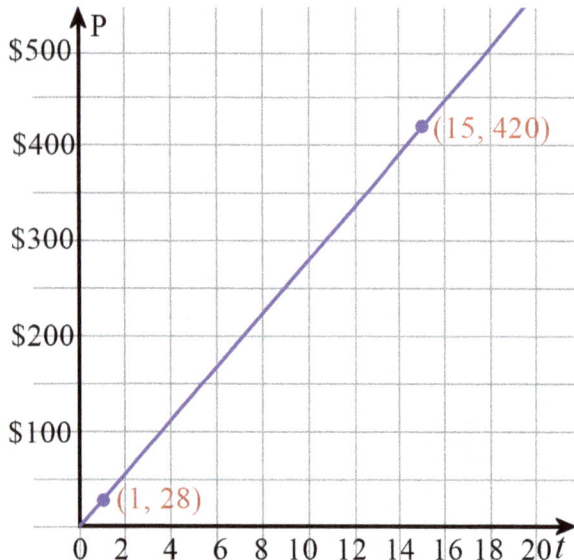

4.

a.	$4x - 8 > -7$	b.	$12 + 5x \leq 50$
	$4x > 1$		$5x \leq 38$
	$x > 1/4$		$x \leq 38/5 = 7\,3/5$

5. a. The volume is $\pi \cdot (9.125 \text{ in})^2 \cdot 27.5 \text{ in} \approx$ <u>7,193 cubic inches</u>.

   b. 31.14 gallons

Skills Review 73, p. 83

1. a. As *x*-values increase by 2, the *y*-values increase by ½. Thus the slope is ½/2 = 1/4.

   b. As *x*-values increase by 1, the *y*-values decrease by 1. (You have to read the table backwards to read the *x*-values in increasing order.) Thus the slope is $-1/1 = -1$.

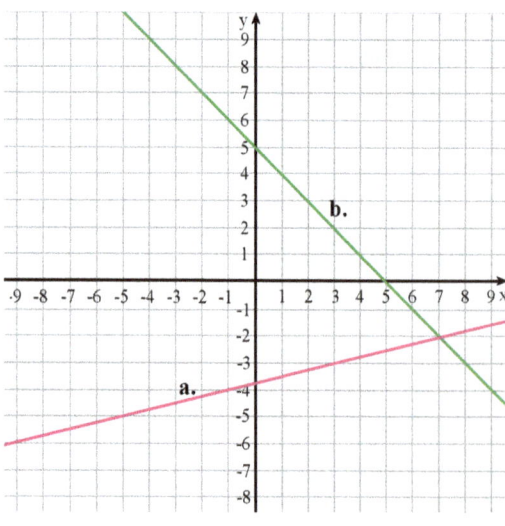

Skills Review 73, cont.

2. The new drawing is 25/15 = 5/3 = $1.\overline{6}$ as long and as high as the other scale drawing.

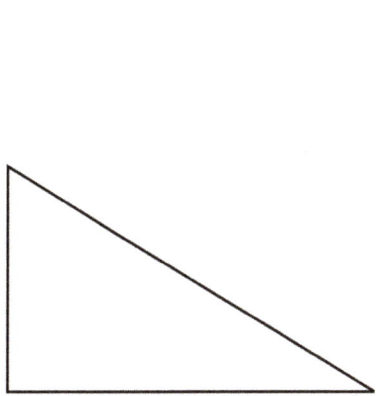

Scale 1 cm = 25 cm   Scale 1 cm = 15 cm

3. a. 3.1623   b. 4.3589   c. 5.2374

4.

a. How much is 62% of 7,300 km?
$\dfrac{x}{7{,}300 \text{ km}} = \dfrac{62}{100}$
$100x = 62 \cdot 7{,}300 \text{ km}$
$x = \dfrac{62 \cdot 7{,}300 \text{ km}}{100}$
$x = 4{,}526 \text{ km}$
b. Forty-three percent of a number is 8.17. What is the number?
$\dfrac{8.17}{x} = \dfrac{43}{100}$
$43x = 817$
$x = 19$

Skills Review 74, p. 85

1.

a.
$$a^2 + 4^2 = 9^2$$
$$a^2 + 16 = 81$$
$$a^2 = 65$$
$$a = \sqrt{65} \approx 8.062$$
$$\text{or } a = -\sqrt{65} \approx -8.062$$

b.
$$36^2 + x^2 = 61^2$$
$$1296 + x^2 = 3{,}721$$
$$x^2 = 2{,}425$$
$$x = \sqrt{2425} \approx 49.244$$
$$\text{or } x = -\sqrt{2425} \approx -49.244$$

2. a. C = 6.4 km · 3.14 ≈ 20.1 km   b. C = 30 cm · 3.14 ≈ 94 cm   c. C = 5.0 in · 3.14 ≈ 15.7 in

3. The Smith family's grocery bill increased by ($412 − $365)/$365 = 47/365 ≈ 12.88%.
   The Daniels family's grocery bill increased by ($345 − $319)/$319 = 26/319 ≈ 8.15%.
   <u>The Smith family's bill increased by the greater percentage</u>.

4.

a. $\dfrac{5^3}{5+2} = \dfrac{125}{7} = 17\,6/7$

b. $\dfrac{12+3}{12-3} = \dfrac{15}{9} = 1\,2/3$

Chapter 10: Probability

Skills Review 75, p. 86

1. a. Yes, because they fulfill the Pythagorean Theorem:

$$16^2 + 12^2 \stackrel{?}{=} 20^2$$
$$256 + 144 \stackrel{?}{=} 400$$
$$400 = 400$$

b. No, they don't. They don't fulfill the Pythagorean Theorem:

$$8^2 + 5^2 \stackrel{?}{=} 11^2$$
$$64 + 25 \stackrel{?}{=} 121$$
$$89 \neq 121$$

2. Seacoast City has $(453{,}295 - 387{,}270)/387{,}270 \approx \underline{17.0\%}$ more inhabitants than Snowflake City.

3. a. 80 km/h

   b. See the image on the right.

4. a. The area is $\pi \cdot (31.5 \text{ cm})^2 \approx 3120 \text{ cm}^2$
   b. The area is $\pi \cdot (177 \text{ in})^2 \approx 98{,}400 \text{ in}^2$

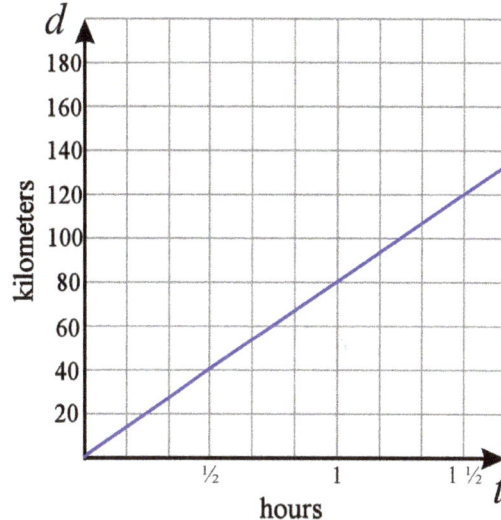

Skills Review 76, p. 87

1. The side of the square is 15 m. We can then solve the diagonal using the Pythagorean Theorem. Let d be the diagonal. Then:

$$15^2 + 15^2 = d^2$$
$$225 + 225 = d^2$$
$$d = \sqrt{450} \approx 21.21 \text{ m}$$

   The diagonal is about <u>21.21 meters</u>.

2. $y = (¼)x - 2$

3. The area is (6 cm + 12 cm)/2 · 7 cm = <u>63 cm^2</u>.

4. The dimensions of this room in reality are 6 ½ in · 2 ft/1 in = 13 ft, and 5 in · 2 ft/1 in = 10 ft.

   a. They would be 13 ft · 1 in/5 ft = 13/5 in = <u>2 3/5 in</u>, and 10 ft · 1 in/5 ft = <u>2 in</u>.

   b. They would be 13 ft · 1 in/3 ft = 13/3 in = <u>4 1/3 in</u>, and 10 ft · 1 in/3 ft = <u>3 1/3 in</u>.

   These could also be solved by multiplying the dimensions 6 ½ in. and 5 in. by a scaling factor, 2/5 in part (a), and 2/3 in part (b).

5. Answers will vary. Check the student's answer. These are the EXACT fractions and percentages portrayed by the circles:

   a. 1/3, 1/6, 1/5, 3/10 or 33.3% 16.7%, 20%, 30%.

   b. 1/8, 1/4, 1/8, 2/5, 1/10 or 12.5%, 25%, 12.5%, 40%, 10%.

   c. 1/4, 1/3, 1/12, 1/3 or 25%, 33.3%, 8.3%, 33.3%.

Skills Review 77, p. 88

1. a. 5.099   b. 13   c. 11.489

2. a. triangular prism   b. square pyramid   c. cone

3. The area of the faces that face the viewer are is 16 m · 7 m + 9 m · 6 m = 112 m^2 + 54 m^2 = 166 m^2.

   The area of the faces facing away from the viewer is the same, 166 m^2.

   The area of the top faces is: 10 m · 9 m + 6 m · 9 m = 90 m^2 + 54 m^2 = 144 m^2.

   The area of the face facing right is 16 m · 9 m = 144 m^2.

   The area of the faces facing left is 7 m · 9 m + 9 m · 9 m = 63 m^2 + 81 m^2 = 144 m^2.

   The total surface area is 166 m^2 + 166 m^2 + 144 m^2 + 144 m^2 + 144 m^2 = <u>764 m^2</u>.

4. a. The possible outcomes are to get letter T, H, A, N, K, S, G, I, or V.

   b. P(K) = 1/12

   c. P(I or G) = 4/12 = 1/3

   d. Answers will vary. Check the student's answer. For example: You get a consonant (probability 9/12 or 3/4). Or, you get a letter that is NOT in the word "GRUMPY" (probability 10/12 or 5/6).

Skills Review 78, p. 89

1.

a.	$4x^2$	=	3,600	b.	$y^2 + 75$	=	700
	x^2	=	900		y^2	=	625
	x	=	30 or		y	=	25 or
	x	=	−30		y	=	−25

2. a. 9/27 = 1/3   b. 11/27

3. a. 516 square inches   b. 3.58 square feet

4. a. Check the student's answer. The image on the right is NOT to scale but shows the basic shape of the triangle. The student's triangle may be a reflected and/or rotated version of this.

   b. It is 45°.

   c. It is scalene and obtuse.

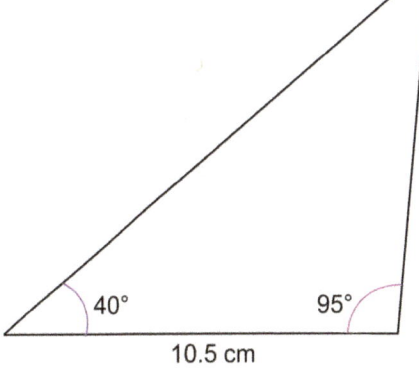

Skills Review 79, p. 90

1. a. Check the student's drawing. There are two possibilities for the orientation. The images below are not to scale.

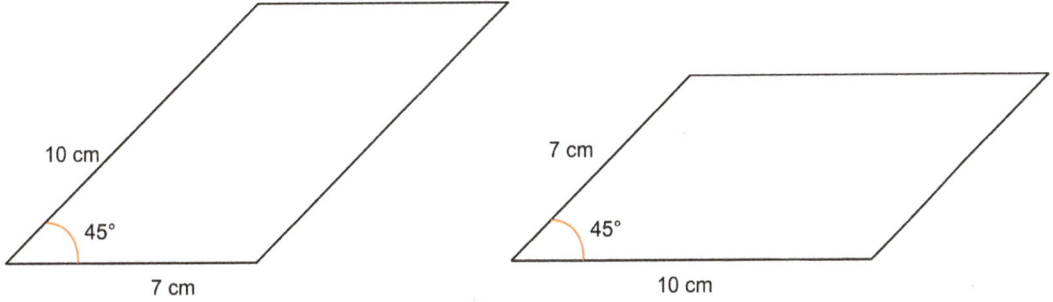

   b. Yes. The two 45° angles on opposite corners of the parallelogram are complementary.
   c. Yes. The two angles on any given side of this parallelogram will be a 45° and 135°, so are supplementary.

2. The dimensions of the park are (approximately):

   The dimensions in reality are 3.3 cm · 30,000 = 99,000 cm = 990 m and 1.7 cm · 30,000 = 51,000 cm = 510 m.
   The area is 990 m · 510 m = 504,900 m² ≈ 505,000 m².
   Since 504,900 m² is 50.49 hectares, the area of the park to the nearest hectare is 50 hectares.

3. a. 7² + 5² = 49 + 25 = 74 and 10² = 100. Since the square of the long side is greater than the sum of the squares of the two shorter sides, these lengths form an obtuse triangle.

   b. 12² + 9² = 144 + 81 = 225 and 15² = 225. Since the square of the long side is equal the to sum of the squares of the two shorter sides, these lengths form a right triangle.

Skills Review 79, cont.

4. Answers will vary a lot. Check the student's answer. Here is one possible result for pulling out socks from a bag that contained two times as many white socks as blue socks.

Outcome	Relative Frequency	Experimental probability (%)
white sock	59/100	59%
blue sock	41/100	41%

Skills Review 80, p. 91

1. a. Rewrite this scale in the format 1 cm = __400__ m.

   b. See the table on the right.

on the map (cm)	in reality
0.875 cm	350 m
2 cm	800 m
3.5 cm	1.4 km
6.75 cm	2.7 km

2. Let s be the number of scarves.
   The inequality is $2 \cdot 18.99 + 7.99s \leq 75$.
   Solution:

$$2 \cdot 18.99 + 7.99s \leq 75$$
$$37.98 + 7.99s \leq 75$$
$$7.99s \leq 37.02$$
$$s \leq 4.63$$

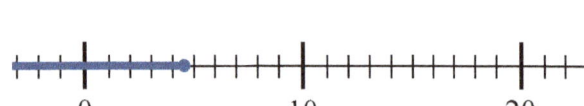

She can buy at most 4 scarves.

3. (a) is set up correctly. In (b), you end up multiplying the dollar amount by the dollar amount, which won't work.

   a. $\dfrac{14 \text{ lb}}{\$91} = \dfrac{x}{\$143}$

   $\$91x = 14 \text{ lb} \cdot \$143$

   $x = \dfrac{14 \text{ lb} \cdot \$143}{\$91}$

   $x = 22 \text{ lb}$

4. a. ∠A = 126°   ∠B = 54°   b. ∠A = 61°

Chapter 11: Statistics

Skills Review 81, p. 92

1. a. See the image on the right. It is not to true scale so check the sides of the student's triangle that they are the correct lengths.
   b. It is obtuse and scalene.

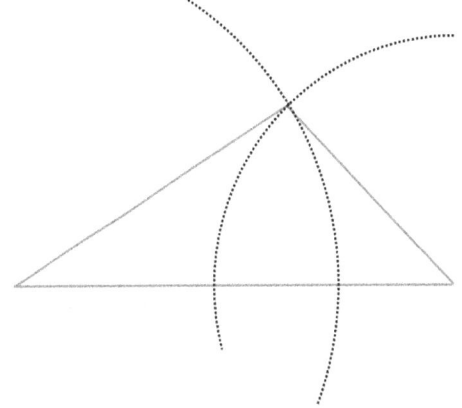

2. Caleb will pay I = Prt = $925 · 0.114 · 8/12 = $70.30 as interest.

   In the other situation, he would have paid
   I = Prt = $925 · 0.107 · 5/12 = $41.24 as interest.

   In the second situation, he would have paid
   $70.30 − $41.24 = $29.06 less.

3. a. 21/26. There are five vowels (A E I O U) and 26 letters, which means there are 21 consonants.
   b. 15/26 (The letters with no round parts are A E F H I K L M N T V W X Y Z.)

4. Let h be the height of the triangle. From the Pythagorean Theorem we get:

$$h^2 + 8.5^2 = 17^2$$
$$h^2 + 72.25 = 289$$
$$h = \sqrt{289 - 72.25}$$
$$h \approx 14.7224$$

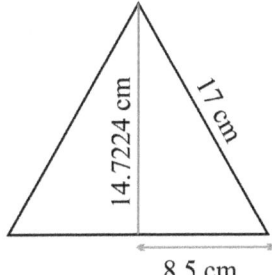

Then, the area of the triangle is 17 cm · 14.7224 cm / 2 ≈ 125 cm².

Skills Review 82, p. 93

1. a. 5.268   b. 60.147

2. a. Answers will vary. Check the student's answer. For example:

   Generate a set of six random numbers, where each number is one of 0, 1, 2, or 3.
   Or: Use a deck of cards so that each of the four suits corresponds to a certain fruit. Draw a random card and put it back. Repeat this six times to get a set of six "fruits".

   b. Answers will vary. Check the student's answer. If you use a random number generator or generate random numbers in a spreadsheet program, you should have 100 rows (or columns) of four numbers, for a total of 400 numbers.

   c. d. e. Answers will vary a lot. For example, for a set of 100 rows of four numbers I (Maria Miller) generated, I got these answers:

   c. P(none are raisins) = 35/100

   d. P(exactly 2 are banana chips) = 20/100
   e. P(2 are pineapple, 1 is a banana chip, and 1 is an apricot) = 6/100

Skills Review 83, p. 94

1. a. $V = A_b h = \pi \cdot (3.15 \text{ in})^2 \cdot 11.8 \text{ in} \approx 368$ cubic inches.
   b. 1.59 gallons

2.

	Circle A	Circle B	Circle C
Circumference	16 cm	11.3 in	57.1 m
Diameter	5.1 cm	3.6 in	18.2 m
Radius	2.5 cm	1.8 in	9.1 m

3.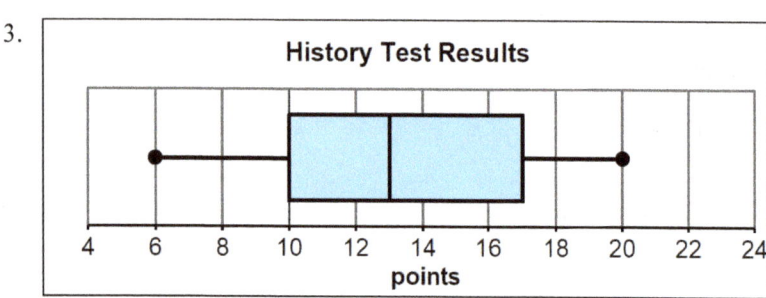

Five-number summary
Minimum 6
First quartile 10
Median 13
Third quartile 17
Maximum 20
Interquartile range 7

4.
a.
$$74 - x^2 = 23$$
$$-x^2 = -51$$
$$x^2 = 51$$
$$x = \sqrt{51} \approx 7.141$$
$$\text{or } x = -\sqrt{51} \approx -7.141$$

b.
$$119^2 + s^2 = 15{,}800$$
$$14{,}161 + s^2 = 15{,}800$$
$$s^2 = 1{,}639$$
$$s = \sqrt{1{,}639} \approx 40.485$$
$$\text{or } s = -\sqrt{1{,}639} \approx -40.485$$

Skills Review 84, p. 95

1.
A rectangle with 0.6 mi and 1.4 mi sides

A = __0.84__ sq mi

A = __0.84__ sq mi · $\dfrac{640 \text{ acres}}{1 \text{ sq mi}}$ = __540__ acres (*to the nearest ten acres*)

2. c. $\pi \cdot 18$

3. a. 20/80 = 1/4   b. 75/80 = 15/16

4. a. Method (5)
   b. Methods (2) and (3)
   c. Method (4)

Skills Review 85, p. 96

1. Answers will vary a lot, since the composition of the group of objects will vary and since the exercise involves a chance process. Check the student's results. See the table on the right for an example.

    The theoretical and experimental probabilities should be relatively close. However, they might not be if the process for drawing the objects was not truly random. Each object has to have an equal chance of being drawn.

Outcome	Theoretical probability	Relative Frequency	Experimental probability
Blue	30%	25/100	25%
Green	20%	21/100	21%
Red	10%	13/100	13%
Yellow	20%	20/100	20%
Pink	20%	21/100	21%

2. a. $V = \pi \cdot (1.4 \text{ in})^2 \cdot 4.0 \text{ in} \approx \underline{25 \text{ in}^3}$.   b. 4.3 cm · 5.0 cm / 2 · 12.0 cm ≈ $\underline{130 \text{ cm}^3}$

Skills Review 86, p. 97

1. The scale factor is 27/24 = 9/8 = 1.125.
   The base side is the longest (see the image in the exercise) and thus it becomes 1.125 · 69 mm = 77.625 mm.
   The area of the enlarged triangle is therefore 77.625 mm · 27 mm / 2 = 1047.9375 m² ≈ $\underline{1048 \text{ mm}^2}$.

2. Using the Pythagorean Theorem, we get:

$$\begin{aligned} h^2 + 3^2 &= 12^2 \\ h^2 + 9 &= 144 \\ h^2 &= 135 \\ h &= \sqrt{135} \approx 11.62 \text{ ft} \end{aligned}$$

The ladder is at the height of about 11.6 ft.

3. a.

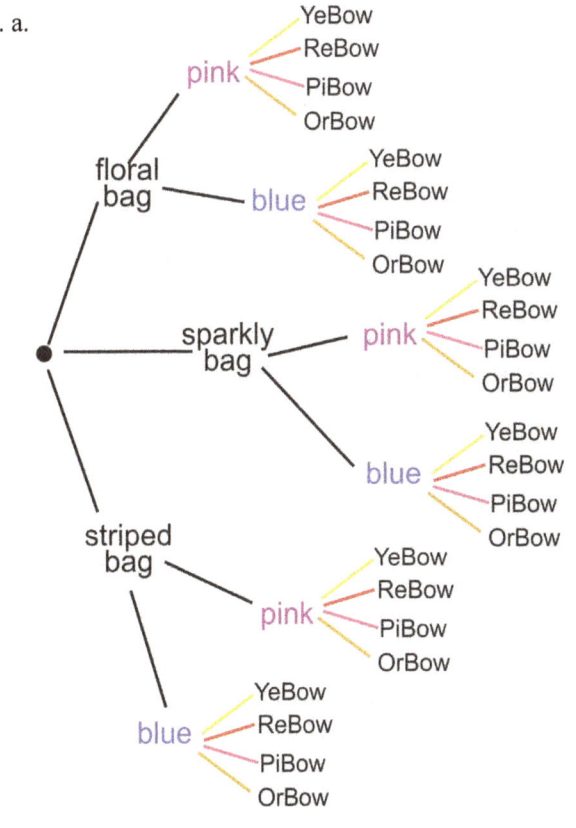

3. b. We can see there are 24 possibilities in total. The probability of this exact combination is 1/24. You can also calculate it as 1/3 · 1/2 · 1/4 = 1/24.

   c. From the tree diagram, we can count that there are 8 favorable outcomes to this, so the probability is 8/24 or 1/3. You can also calculate this as 1/2 · 2/3 = 1/3.

   d. Half of the possibilities have a pink or red bow (neither yellow nor orange), so the probability is 1/2.

   e. There are four favorable outcomes, so the probability is 4/24 = 1/6. You can also calculate this as 1/3 · 2/4 = 1/6.

Skills Review 87, p. 98

1. a.

Farm 1 Weights (grams)	
49	52
49	52
49	52
49	52
50	52
50	53
50	53
50	54
50	55
50	55
50	55
51	55
51	56
51	56
52	56

<u>Five-number summary</u>

Minimum: <u>49</u>

1st quartile: <u>50</u>

Median: <u>52</u>

3rd quartile: <u>54</u>

Maximum: 56

Interquartile range: <u>4</u>

Farm 2 Weights (grams)	
53	57
54	57
54	57
54	57
54	57
54	57
55	58
55	58
55	58
55	58
55	59
55	59
55	60
56	60
56	60

<u>Five-number summary</u>

Minimum: <u>53</u>

1st quartile: <u>55</u>

Median: <u>56.5</u>

3rd quartile: <u>58</u>

Maximum: <u>60</u>

Interquartile range: <u>3</u>

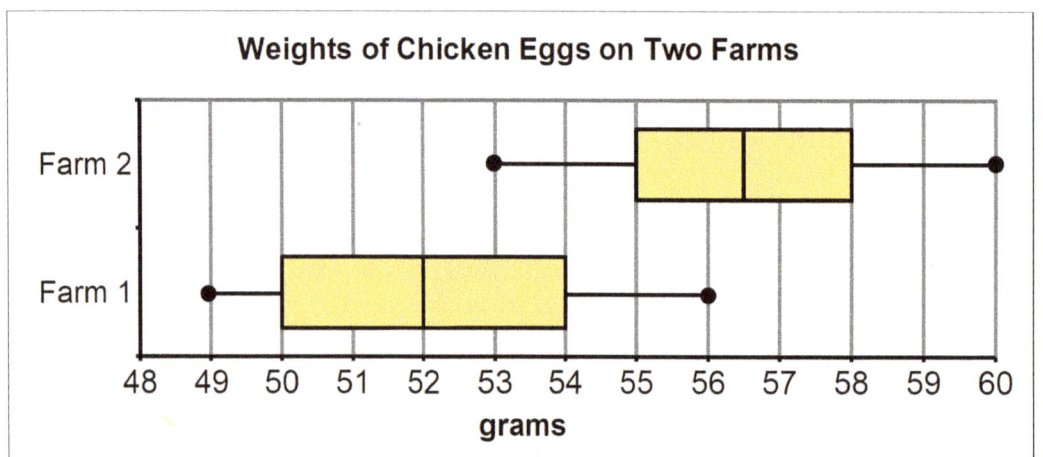

b. The distributions overlap about half-way, or partially, or somewhat, from 53 grams to 56 grams. (The extent of the overlap, 3 grams, is close to half of the range of either distribution.)

c. Farm 2.

d. Yes, it is significant. The difference in the medians is 4.5 grams. The interquartile ranges are 3 and 4. So, the difference between the medians is about one time the measure of variability and that makes it significant.

Skills Review 88, p. 99

1. a. <u>Group 2</u> appears to spend more hours on social media.

   <u>Neither group</u> appears to have greater variability (just looking at the distribution, without calculating anything).

   b. <u>Group 1</u>:

   Median <u>30</u>   Range <u>27</u>

   Interquartile range: 6

   <u>Group 2</u>:

   Median <u>36</u>   Range <u>27</u>

   Interquartile range: 7

Number of Hours Spent Using Social Media in One Month		
Group 1 Leaf	Stem	Group 2 Leaf
99988	1	
99998764	2	445899
433322110000	3	033344556677789
53110	4	002488
	5	01

   c. Do these values support your answers in (a)?

   Yes. The difference in the medians is 36 − 30 = <u>6</u>. This is about one time the interquartile range. So, the difference in the medians is significant, which supports our notion that just by looking a the plot alone you can see that group 2 seems to spend more time on social media than group 1.

   Also, both the ranges and the interquartile ranges are similar for both groups, which supports our notion from the graph that neither groups appears to have greater variability in the data.

Skills Review 89, p. 100

1. The area in square meters is 3 m · 3 m = 9 m^2.

   The area in square *decimeters* is <u>30</u> dm · <u>30</u> dm = <u>900</u> dm^2.

   The area in square *centimeters* is <u>300</u> cm · <u>300</u> cm = <u>90,000</u> cm^2.

   The area in square *millimeters* is <u>3000</u> mm · <u>3000</u> mm = <u>9,000,000</u> mm^2.

   So, 9 m^2 = <u>900</u> dm^2 = <u>90,000</u> cm^2 = <u>9,000,000</u> mm^2.

2. a. See the image on the right (not to scale):

   b. Each triangular area is 8 cm · 6.5 cm ÷ 2 = 26 cm^2.
   The square itself is 64 cm^2.
   So the total surface area is 4 · 26 cm^2 + 64 cm^2 = 168 cm^2.

3. Answers will vary. Check the student's answer. For example, we can draw these conclusions:

   - Almond Lavender and Sweet Almond are the most popular oils.

   - Coconut and Jojoba Rose are not nearly as popular as Almond Lavender and Sweet Almond oils. Looking at the numbers, they are about half as popular as Almond Lavender and Sweet Almond.

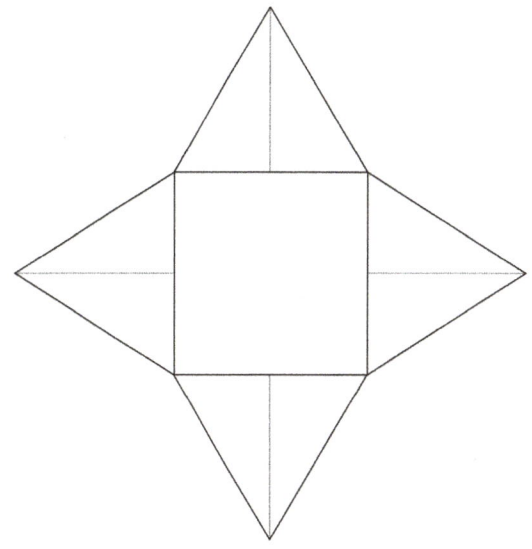

Skills Review 90, p. 101

1. Answers will vary. Check the student's answer.

   For example: Make up a deck of cards so that 30% of them are hearts. For example, make a deck with 12 hearts and 28 other cards. Then choose a card randomly, and put it back. Repeat this eight times.

   Or, you can generate random integers between 0 and 9 such as at https://random.org/inters or using a spreadsheet. To get the 30% probability, let the integers from 0 to 2 represent coming to work early and the rest to represent not coming to work early.

   In this situation, generating 800 integers and formatting the results in 8 columns will get you 100 rows with eight numbers in each. Each row of eight numbers represents one repetition of choosing of 8 workers randomly.

   Here is an example result of the simulation:

Results of simulation		
Employees who arrived at work early	Relative Frequency	Experimental Probability
0	5/100	5%
1	16/100	16%
2	28/100	28%
3	25/100	25%
4	19/100	19%
5	6/100	6%
6	1/100	1%
7	0/100	0%
8	0/100	0%
TOTALS		100%

   Answers to the questions will vary based on the results of the student's simulation.

   a. What is the probability that two of the eight employees have arrived at work early? 28%

   b. What is the probability that only one employee has arrived at work early? 16%

   c. What is the probability that none of the eight arrived at work early? 5%

   d. What is the probability that *at most* 2 of them have arrived at work early? 49%

   e. What is the probability that *at least* 3 of them have arrived at work early? 51%

www.ingramcontent.com/pod-product-compliance
Lightning Source LLC
Chambersburg PA
CBHW081349040426
42450CB00015B/3364